Water Perspectives in Emerging Countries

Focus Issue

Microplastics in the Water Environment

Sandhya Babel, Andreas Haarstrick, Mukand S. Babel , Alice Sharp

August 19 - 21, 2019 - Island of Koh Samui, Thailand

Water Perspectives in Emerging Countries

Focus Issue

Microplastics in the Water Environment

Sandhya Babel, Andreas Haarstrick, Mukand S. Babel, Alice Sharp

August 19 - 21, 2019 - Island of Koh Samui, Thailand

Issue Editors
Prof. Dr. Sandhya Babel
Thammasat University, School of Biochemical Engineering and Technology, Sirindhorn International Institute of Technology, PO Box 22, Pathumthani 12121, Thailand; sandhya@siit.tu.ac.th

Prof. Dr. Andreas Haarstrick
Technische Universität Braunschweig, Leichtweiss Institute, Exceed Office, Beethovenstr. 51 B, 38106 Braunschweig, Germany; a.haarstrick@tu-bs.de

Prof. Dr. Mukand S. Babel
Asian Institute of Technology, Water Engineering and Management, Chair Climate Change Asia at AIT, PO Box 4, Klong Luang, Pathumthani 12120, Thailand; msbabel@ait.ac.th

Dr. Alice Sharp
Chiang Mai University, Faculty of Science, Environmental Science Research Centre, 239 Huaykaew Road, Tumbol Suthep, Amphur Muang, Chiang Mai 50200, Thailand; alice.sharp@cmu.ac.th

Exceed Chairman & Editor-in-Chief
Prof. Dr.-Ing. Norbert Dichtl
Technische Universität Braunschweig, Institute of Sanitary and Environmental Engineering, Pockelsstr. 2a, 38106 Braunschweig, Germany; n.dichtl@tu-bs.de

Publishing Editor
Prof. em. Dr. mult. Dr. h.c. Müfit Bahadir
Technische Universität Braunschweig, Leichtweiss Institute, Exceed Office, Beethovenstr. 51 B, 38106 Braunschweig, Germany; m.bahadir@tu-bs.de

This publication was financed by the German Academic Exchange Service (DAAD) and the Federal Ministry for Economic Cooperation and Development (BMZ).
All rights reserved including translation into foreign languages. The publication or parts thereof may not be reproduced in any form without permission from the publishers.
Printed in Germany by Cuvillier Verlag, Göttingen, Germany

Bibliografische Information der Deutschen Nationalbibliothek
Die Deutsche Nationalbibliothek verzeichnet diese Publikation in der
Deutschen Nationalbibliografie; detaillierte bibliografische Daten sind im Internet
über http://dnb.d-nb.de abrufbar.
1. Aufl. - Göttingen: Cuvillier, 2019

© CUVILLIER VERLAG, Göttingen 2019
    Nonnenstieg 8, 37075 Göttingen
    Telefon: 0551-54724-0
    Telefax: 0551-54724-21
    www.cuvillier.de

Alle Rechte vorbehalten. Ohne ausdrückliche Genehmigung des Verlages ist
es nicht gestattet, das Buch oder Teile daraus auf fotomechanischem Weg
(Fotokopie, Mikrokopie) zu vervielfältigen.
1. Auflage, 2019
Gedruckt auf umweltfreundlichem, säurefreiem Papier aus nachhaltiger Forstwirtschaft.

  ISBN 978-3-7369-7089-2
  eISBN 978-3-7369-6089-3

CONTENT

Preface	1
Current status of microplastics contamination in marine and freshwater environments (A.T. Ta, S. Babel)	2
Recent advances in microplastic research in Japan (A. Isobe)	14
Presence and distribution of microplastics from upstream to downstream of Citarum River, West Java, Indonesia (E. Sembiring, F.C. Alam, V. Suendo, M. Reza)..	24
Prospects of microplastics in the sedimentary deposits on Patos Lagoon Coast (E.S. Vogelmann, V.R. Alegrini, G.O. Awe, J. Prevedello)	34
Sampling strategy to conduct studies on spatial and temporal contamination of microplastics in estuarine ecosystems (M. Barletta)	43
Detection and identification of microplastic materials waste in water systems (S. El Hajjaji, J. Mabrouki, C. Bakkouche, A. Dahchour)	54
Influence of typical ions and natural organic matter on the aggregation of microplastic particles in aqueous phase (R. Liu, H. Wang)	63
Microplastics in human consumption: table salt contaminated with microplastics (E.V. Ramasamy, S.N. Sruthi, A.K. Harit, N. Babu)	74
Microplastics removal from municipal wastewater by an electrocoagulation system (G.H. Bracher, C. Graepin, D. Elkhatib, V. Oyanedel-Craver, E. Carissimi)	81
Potential of plastic-degrading microbe in preventing plastic pollution and its circular sustainability program (D.K.T. Sukmadewi, A.B. Mustafa)	92
Microplastics challenges and international governance mechanisms (M. Mumtaz, J.A. Puppim de Oliveira, I. Ahmed)	102
An evaluation of downstream policies to reduce marine plastic litter in Thailand (S.K.M. Huno, G. Borongan, N. Tsukamoto)	115

PREFACE

Today, plastic has become virtually indispensable in our daily lives. The wide plastic family is composed of a variety of materials designed to meet different needs of thousands of end products. The global plastic production has almost reached 350 million tons in 2017. Based on the current trends, a future scenario predicts that 40 billion tons of plastics will be produced by 2050. At the end, it leads to an ever-increasing quantity of plastic waste, much of which ends up in marine and freshwater environments. The long-term persistence of plastics in the water environment results in microplastics (particles below 5 mm), which are recognized globally as an emerging contaminant. Microplastic contamination in marine and freshwater bodies causes a lot of undesirable impacts, collectively on the global environment and humans. Not only the ecological hazards, but chemical and biological hazards are also associated with microplastics. A large number of additives incorporated in the polymer blend are leached out from microplastics, potentially causing toxic effects on both humans and aquatic organisms, if ingested. On the other hand, microplastics are efficient vectors that transfer heavy metals and micro-organisms bound to them over long distances.

Most of the microplastics studies have been conducted in Europe and United States, and very few investigations in Asian and African regions. Still many gaps exist in microplastics research, e.g., sources of origin, their fate inside human and organism bodies, and subsequent health implications. Researchers have largely focused on marine microplastics as the major portion of the plastic debris ends up in oceans. But freshwater ecosystems are exposed to microplastics often due to the greater proximity to densely populated areas and receiving discharges from wastewater treatment plant. Now, microplastics are reported to be found in drinking water, salt, beer, honey, etc. So, we will soon encounter its dangers. Therefore, it is an hourly need to discuss and investigate on this topic.

The Expert workshop on *"Microplastics in the Water Environment"* was held at the Island of Koh Samui, Thailand, August 18 – 22, 2019. The main objectives of the workshop were to provide the participants with state-of-the-art knowledge of microplastic pollution and its related implications, and secondly, identifying the research needs and concepts on the increasing threats to ecosystems, wildlife and human health. Selected papers from this workshop are presented in this book. This book will provide readers with an improved understanding of microplastics genesis and its threat, bear a holistic view on the microplastics problem and its inherent risk potential.

Prof. Dr. Sandhya Babel, co-editor
School of Biochemical Engineering and Technology
Thammasat University, Sirindhorn International Institute of Technology
Pathumthani 12121, Thailand

CURRENT STATUS OF MICROPLASTICS CONTAMINATION IN MARINE AND FRESHWATER ENVIRONMENTS

Anh Tuan Ta, Sandhya Babel

School of Biochemical Engineering and Technology, Sirindhorn International Institute of Technology, Thammasat University, P.O. Box 22, Pathum Thani 12121, Thailand; sandhya@siit.tu.ac.th

Keywords: Microplastics, sources, freshwater pollution, marine debris, plastic waste

Abstract

Microplastics (MPs) are ubiquitously found all over the world, from densely populated areas to remote ones. The presence of MPs poses a great threat to the entire ecosystem. The pollutants mainly come from daily use products such as cosmetics, paints, etc. (primary MPs); or the breakdown of larger macroplastic debris under environmental conditions (secondary MPs). MPs enter aquatic environments through terrestrial and land-based activities. Since the last few decades, about 93-236 thousand metric tonnes of MP particles have accumulated in the marine environment due to improper disposal of waste plastics. With this situation, this paper reviews sources of MPs and the existing state of MPs in the marine and freshwater environments. In marine environment, MPs are found in their highest concentrations along coastlines and within mid-ocean gyres. Fiber and fragment are the majority morphologies of MPs found in oceans. While a large number of studies on MPs have focused on the marine environment, few are reported in the freshwater environment.

1 Introduction

The plastic waste poses a serious threat to the aquatic wildlife and ecosystem. The Convention on Biological Diversity reported that currently 663 species of aquatic biota are known to be impacted by plastic pollution through entanglement or ingestion, including mammals, birds, and reptiles [1]. Ingested plastics can cause internal damage, reduce feeding, disturb the digestive enzyme system or hormone balance and have an impact on reproduction. In recent years, a particular concern is the occurrence of smaller pieces of plastic debris including those not visible to the naked eye, referred to as MPs. According to 5 Gyres Institute, it is estimated that a total of 15-51 trillion MP particles have accumulated in the ocean, weighing between 93 and 236 thousand metric tons [2].

The term MPs has been defined differently by various researchers. Gregory and Andrady [3] defined MPs as the barely visible particles that pass through a 500 μm sieve but retained on a 67 μm sieve (0.06 – 0.5 mm in diameter), and particles larger than this were called mesoplastics, while others as Fendall and Sewell [4] and Moore [5] defined the MPs as being in the size range < 5 mm (recognizing 333 μm as a practical lower limit, when neuston nets are used for sampling). This is in line with the definition of the European Commission in the Marine Strategy Framework Directive (2008/56/EG), particles < 5 mm are considered as MPs [6]. In addition, the MPs have the following properties: (1) Solid phase materials, (2) Insoluble in water, (3) Synthetic, (4) Non-degradable, (5) Made from plastic.

Based on the above facts, main goal of this paper is to create an overview about current situations of MP pollution in marine and fresh water environments.

2 Sources of MPs

MPs originate from a variety of sources, but these can be categorized as primary and secondary sources. These pollutants get washed down the sink, as they are too small to be filtered by sewage-treatment plants, consequently ending up in the fresh systems (i.e., rivers, lakes, or canals) and ultimately in the oceans.

2.1 Primary source

The primary sources of MPs include microbeads in cleaning and cosmetic, as shown in Figure 1, or manufactured pellets used in feedstock or plastic production. The plastic materials applied as ingredients in cosmetic formulations include the two main categories of plastics typically made from petroleum carbon sources. Thermoplastics include polyethylene (PE), polypropylene (PP), polystyrene (PS), polytetrafluoroethylene (Teflon), and polymethyl methacrylate, while thermoset plastics include e.g., polyesters, polyurethanes [7]. Based on the usage of PE microbeads in personal care products, Gouin et al. [8] estimated that the U.S. population may be emitting about 263 t/yr of PE microbeads or approximately 2.4 mg per person per day.

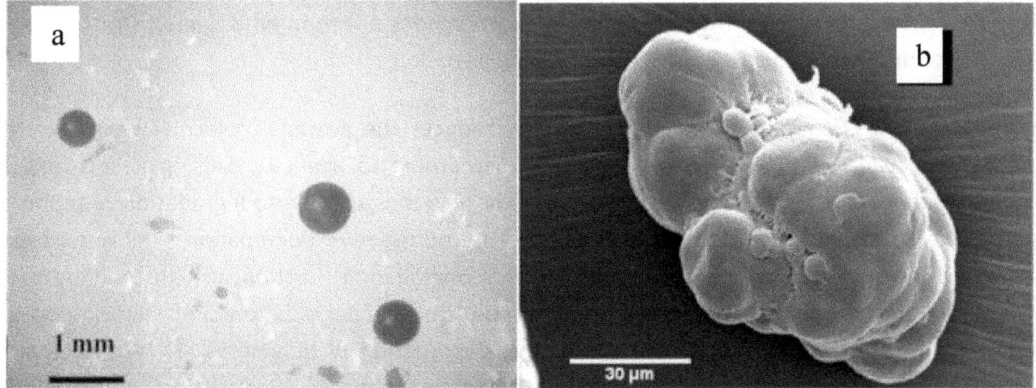

Figure 1: (a) Photomicrographs of a microbead sample from the sea surface of Lantau Island, Hongkong; (b) Scanning electron microscopy (SEM) of a typical rough facial scrub plastic microbead particle [9]

Plastic resin pellets or flakes and plastic powder or fluff are another important source of primary MPs. Plastic pellets (around 5 mm diameter) and powders (less than 0.5 mm) are generally with the shape of a cylinder or a disk [10]. These plastic particles are the industrial feed stock for production of plastic products and transported to manufacturing sites, where they are re-melted and molded into a wide range of final products. They can be unintentionally released to the environment after an accidental loss during transport or with run-off from processing facilities, i.e. often as a result of improper handling.

2.2 Secondary source

The secondary sources of MPs include fibers from textile laundering or fragments from the breakdown of larger plastic items. Large plastic debris on sea and land degrades over time into smaller particles until they end up as MPs when exposed to some factors in the environments (Figure 2). This is the most likely process for the generation of secondary MPs in the marine environment [11]. It is important to note that the formation of MPs is influenced by a combination of environmental factors and the properties of the polymer.

Figure 2: (a) Degradation and fragmentation of plastic under environmental factor; (b) the cracks seen at the surface are caused by photochemical degradation

Degradation is a chemical change that drastically reduces the average molecular weight of the polymer [12]. Degradation is generally classified according to agencies causing it such as: (a) Biodegradation – action of living organisms usually microbes; (b) Photodegradation – action of light (usually sunlight in outdoor exposure); (c) Thermooxidative degradation – slow oxidative breakdown at moderate temperatures; (d) Thermal degradation – action of high temperatures; and (e) Hydrolysis – reaction with water. The dominant cause of degradation of plastics outdoors is solar UV radiation, which facilitates the oxidative degradation of polymers [13]. With common polymers such as LDPE, HDPE, PP and nylons exposed to the marine environment, it is primarily the UV-B radiation in sunlight that initiates photo-oxidative degradation. Once initiated, the degradation can also proceed thermooxidatively without the need for further exposure to UV radiation. The autocatalytic degradation reaction sequence can progress as long as oxygen is available to the system [12]. During advanced stages of degradation, the plastic debris typically discolors, develops surface features, becoming weak and brittle (embrittle) in consequence over time. Any mechanical force (e.g., wind, wave, animal bite and human activity) can break the highly degraded, embrittled plastics into fragments.

Fibers from synthetic textiles are one of the sources to secondary MPs in the environments (Figure 3). Synthetic fibers are produced from crude oil through polymerization, polycondensation or polyaddition processes [14]. Textiles made from synthetic materials commonly consist either of fibers made out of long filaments or of fibers that have been cut into shorter fibers. In 1950, there was an annual production of 2.1 million tons of synthetic fibers. With increased demand, the production reached 49.6 million tons in 2010 [15]. Textile synthetic fibers are not only used in clothes but also in furniture, geotextiles, cloth, footballs, backpacks, cuddly toys, buildings, and agriculture.

Different kinds of fabrics have different abilities to shed fibers. The "sheddability" depends on the fabric type, the texture, the yarn type, the nature and the number of the fiber types involved. It also depends on whether the fabric is made out of staple fibers or filaments. According to Astrom [14], most fibers are lost from synthetic fleece and microfleece, and there was a large difference between the various fleece fabrics and the rest of the fabrics. According to Browne et al. [16], a synthetic fleece jacket can release around 1,900 fibers every wash. The authors also concluded that using detergent results in more shedding compared with using water alone. The size of fibrous plastic particles has a wide range according to different studies. Thompson et al. [17] set the size to 20 µm in diameter, while Napper et al. [18] concluded that average fiber size ranged between 11.9 and 17.7 µm in diameter, and 5.0 and 7.8 mm in length. Compared with other plastics found in oceans, like pellets or nibs, fibers have a greater surface to volume ratio. This could mean that they can attract more chemicals than other MPs.

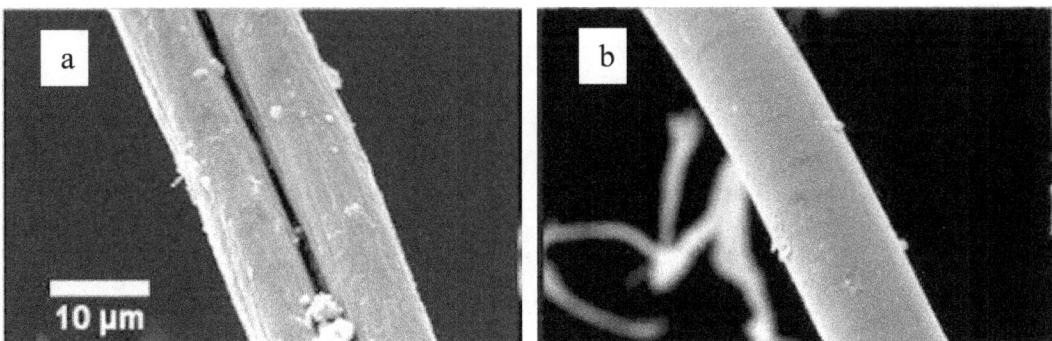

Figure 3: Scanning electron microscopy image (SEM) of typical fibers: (a) Polyester Cotton Blend; (b) Polyester [18]

3 MPs in marine environment

MPs enter the marine environment via different pathways (terrestrial and marine-based activities), as shown in Figure 4. Plastic litters with a terrestrial source contribute ~80% of the plastics found in marine litter [12]. MPs used both in cosmetics and as air-blasting media can enter waterways via domestic or industrial drainage systems [19]. Wastewater treatment plants can trap macro-plastics and some small plastic debris within oxidation ponds or sewage sludge, but a large proportion of MPs will pass through such filtration systems [4, 20]. With approximately half the world's population residing on coasts, these kinds of plastic have a high potential to enter the marine environment via rivers and wastewater systems, or by being blown off-shore [5, 21].

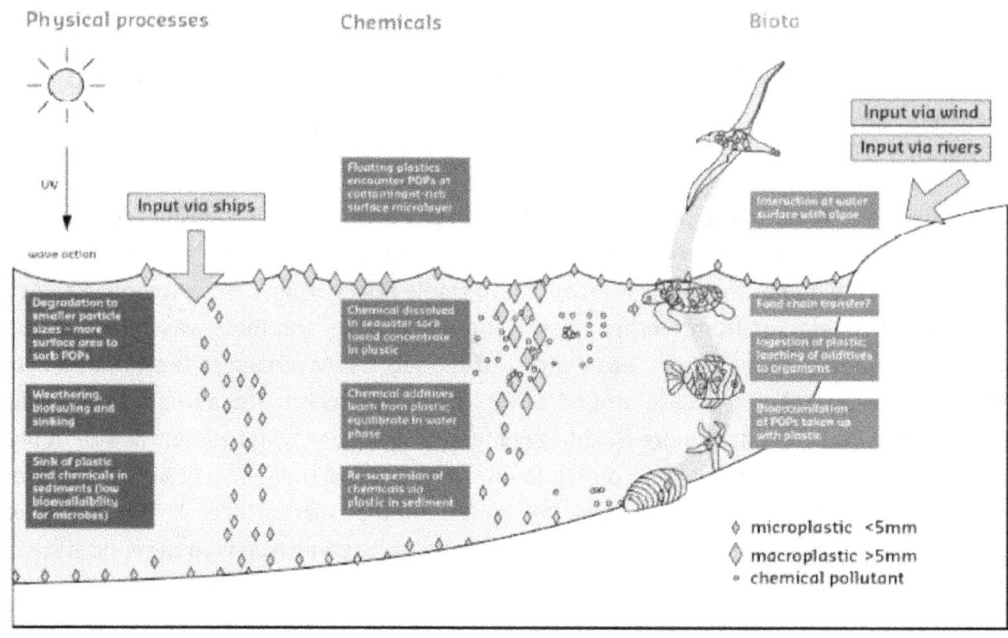

Figure 4: Sources of marine MPs and the various physical, chemical and biological processes affecting MPs in the marine environment [22]

Plastics that enter river systems either directly or with wastewater effluent or in refuse site leachates will then be transported to the sea. Some studies have shown, how the high unidirectional flow of freshwater systems drives the movement of plastic debris into the oceans [23, 24]. Using water samples from two Los Angeles (California, USA) rivers collected in 2004–2005, Moore [5] quantified the number of plastic fragments present that were <5 mm in diameter. Extrapolating the resultant data revealed that these two rivers alone would release over 2 billion plastic particles into the marine environment over 3 days. Extreme weather, such as flash flooding or hurricanes, can exacerbate this transfer of terrestrial debris from land to the sea [25, 26]. Studies conducted by Moore et al. [24] showed that MPs in Californian waters near the mouth of a modified Los Angeles storm water conveyance system increased from 10 plastic items/m^3 to 60 plastic items/m^3 following a storm.

Coastal tourism, recreational and commercial fishing, marine vessels and marine-industries (e.g., aquaculture, oil-rigs) are all sources of plastic that can directly enter the marine environment, posing a risk to biota. Tourism and recreational activities account for an array of plastics being discarded along beaches and coastal resorts [19], although it is worth noting that marine debris observed on beaches will also arise from materials carried on in-shore- and ocean currents [21]. Fishing gear is one of the most commonly noted plastic debris items with a marine source [12].

Discarded or lost fishing gear, including plastic monofilament line and nylon netting, is typically neutrally buoyant and can, therefore, drift at variable depths within the oceans. The distribution of MPs and their concentrations in the marine environment are shown in Table 1.

Table 1: MPs distribution in marine environments

Water body Name and location	References	Max. and Min. abundance	MPs shape and polymer type
Surface sediment Coastal mangrove, Singapore, Asia	Nor and Obbard [27]	Max: 15.7 items/ 250 g dried, Mean: 9.2 items/ 250 g dried Represents size <40 μm: 58%	Fibers 72.0% Films 23.3% Granules 4.7% Polypropylene, polyvinyl chloride, polyethylene, nylon
Surface water Baltic Sea, Stockholm Archipelago, Sweden	Gewert et al. [28]	Mean: 109,800 items/km^2	Polypropylene: 53% Polyethylene: 24%
Surface water Iskenderun Bay and Mersin Bay, Turkey	Gündoğdu & Çevik [29]	Iskenderun Bay 0.2254 item/ m^2 (mean size: 2.77 mm) Mersin Bay 0.6827 item/ m^2 (mean size: 3.01 mm)	Plastic Fragments: 60.1% Plastic Films: 45.9%
Benthic sediment Baltic Sea, South Eastern	Zobkov & Esiukova [30]	Mean: 34 items/ kg dried	Fibers Fragments Films
Sand, Norderney, German East Frisian Islands, Germany	Dekiff et al. [31]	Max: 2.3 items/ kg dried Min: 1.3 items/ kg dired	PP, PE, polyethylene terephthalate, PVC, PS and PA
Sand, Slovenian coast Shoreline and Infralittoral, Slovenia	Laglbauer et al. [32]	Shoreline: Mean: 133 items/ kg dried Infralittoral: Mean: 155.6 items/ kg dried Represents size: 0.25 – 1 mm: 26% 1 – 5 mm: 74%	Shoreline, Fibers: 96% Films: 4% Infralittoral, Fibers: 75% Fragments: 21% Films: 4%

4 MPs in freshwater environment

The body of knowledge on the accumulation and effects of plastics in freshwater and terrestrial systems is much less than in marine systems [33]. In the last few years, studies of MPs in freshwater environments are rapidly advancing, with MP particles found across a range of freshwater environments worldwide, including lakes and rivers. Area of water surface, depth, wind, currents, and density of particles are all factors determining transport and fate of particles within these aquatic systems.

A significant direct input of primary MPs to fresh water environments has been identified as being through the application of sewage sludge containing synthetic fibers or MPs from personal care or household products. Secondary sources of MPs derive from plastic litter during municipal solid waste collection, processing and land-filling. This includes large plastic items and sanitary waste input to rivers via combined sewage overflows. Runoff via drainage ditches from agricultural land or storm drains from roads containing plastics such as tire wear particles, vehicle-derived debris or fragments of road-marking paints is another significant source of riverine MP loads [34]. Additionally, wind action may also transport lighter plastic items into water bodies or across land [35], and there is evidence to suggest that anthropogenic fibers can be transported and deposited by atmospheric fallout. The MP sources and flows throughout terrestrial and freshwater are shown in Figure 5.

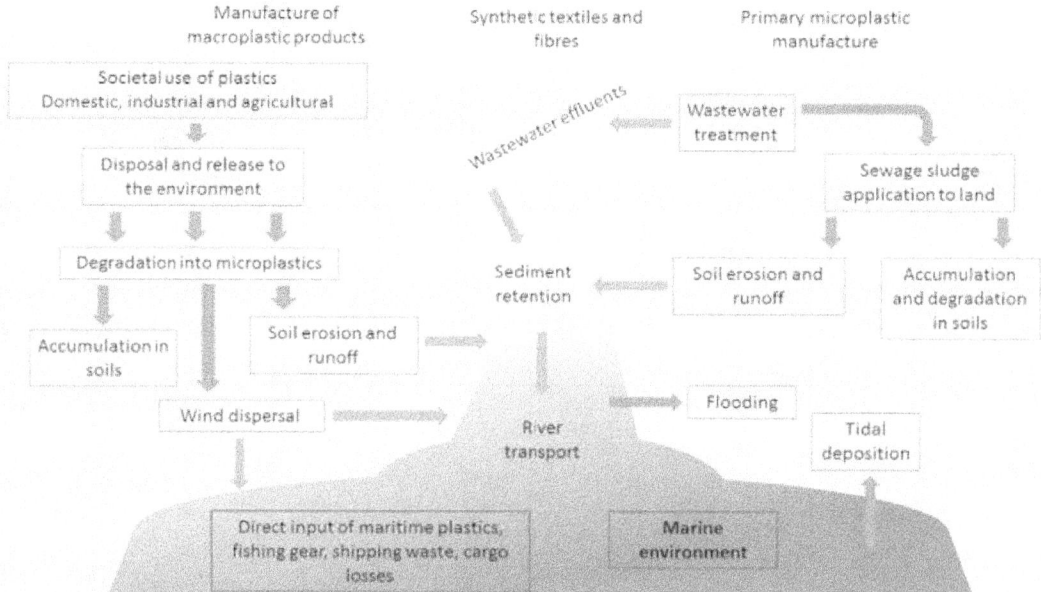

Figure 5: MPs sources and flows throughout terrestrial, freshwater and marine environments [36]

In freshwater environment, MPs have been found in North America, in the Los Angeles basin [37], the North Shore Channel of Chicago [38], Lake Winnipeg [39] and the Great Lakes [34, 40]; in Europe, in Lake Geneva [41], Seine River, Marne River [42], the Austrian Danube River [43], the

Rhine and Main Rivers, Germany [44], and the UK Tamar estuary [45]; and in Asia, in Lake Hovsgol, Mongolia [46], Beijiang River [47], Taihu Lake [48].

According to Free et al. [46], MPs in surface water in Lake Hovsgol, Mongolia in 2014 varied from 20,264 to 44,435 items/km^2. Fragments and films were the most abundant MPs types; no plastic microbeads and few pellets were observed. MPs density decreased with distance from the southwestern shore, the most populated and accessible section of the park, and was distributed by the prevailing winds. Another study by Wang et al. [47] found the concentrations of MPs ranged from 178 ± 69 to 544 ± 107 items/kg sediment. PE, PP, and copolymers were identified in majority. In North America, the average abundance of MPs varied from 43,000 to 466,000 items/km^2 in surface water of the Laurentian Great Lakes. Many MP particles were multi-colored spheres, which were compared to and are suspected to be microbeads from consumer products [34]. The MPs detected in these studies are of varied origins including primary and secondary sources and also different compositions, as shown in Table 2.

Table 2: MPs distribution in fresh water environments

Water body Name and location	References	Max. and Min. abundance	MPs shape and polymer type
Surface water, Lake Hovsgol, Mongolia, Asia	Free et al. [46]	Max: 44,435 items/km^2, Mean: 20,264 items/km^2 Represents size < 4.75 mm: 81%	Fragment: 40% Film: 38% Line/fiber: 20%
Sediment, Beijiang River, China, Asia	Wang et al. [47]	Max: 544 ± 107 items/kg Mean: not indicated Abundance range: 178–554 items/ kg	PE: 42 – 66% PP: 17 – 33 % Copolymers: 6 – 32%
Surface water Laurentian Great Lakes, USA, North America	Eriksen et al. [34]	Max: 466,000 items/ km^2 Mean: 43,000 items/ km^2 Represents size 0.355 – 0.999 mm: 81%	Fragment Pellet
Shore Sediment, Rhine River, Main River, Germany	Klein et al. [44]	Max: 3763 items/ kg Min: 228 items/ kg Max:1368 items/ kg Min: 786 items/ kg Abundance size: 630 – 5,000 μm (weight) 63 – 200 μm (particles)	63 – 200 μm Fibers: 50% Spheres: 13% Fragments: 37% 630 – 5,000 μm Fibers: 13% Spheres: 5% Fragments: 51%

5 Conclusions

MPs are plastic particles smaller than 5 mm that get washed down the sink as they are too small to be filtered by sewage treatment plants, consequently ending up in the river systems and ultimately in the oceans. Currently, large quantities of studies on MPs in the marine environment have been widely carried out, while few were available in the freshwater environment. The effects of MPs on the aquatic environments and humans are still being studied. However, there is much evidence for significant harm to the pollution on wildlife and human health. As shown in the paper, MPs were found in both freshwater and marine environments all over the world from the Northern hemisphere (Europe, North America) to the Southern hemisphere (Asia, Africa). MPs with morphologies of fibers and fragments are dominant in most of the reported studies. Moreover, PP and PE are the most abundant MPs in both fresh and marine environments.

6 Acknowledgements

We would like to thank EXCEED Swindon project and DAAD (German Academic Exchange Service) for support to participate at the Expert Workshop on "Microplastics in the Water Environment", held in Koh Samui, Thailand on August 18-22, 2019. The PhD scholarship was provided to the first author by the Thailand Research Fund (PHD/0241/2560). This study was also partially supported by Asia-Pacific Network for Global Change Research (APN).

7 References

[1] Gall, S.C., Thompson, R.C.: The impact of debris on marine life. Marine Pollution Bulletin, 2015, 92(1-2), 170-179.

[2] Britain, G.: House of Commons Environmental Audit Committee Environmental Crime: Environmental impact of microplastics. 2016: Proquest LLC.

[3] Gregory, M.R. Andrady, A.L.: Plastics in the marine environment. Plastics and the Environment, 2003, 379, 389-90.

[4] Fendall, L.S., Sewell, M.A.: Contributing to marine pollution by washing your face: microplastics in facial cleansers. Marine Pollution Bulletin, 2009, 58(8), 1225-1228.

[5] Moore, C.J.: Synthetic polymers in the marine environment: a rapidly increasing, long-term threat. Environmental Research, 2008, 108(2), 131-139.

[6] Directive, M.S.F.: Directive 2008/56/EC of the European Parliament and of the Council of 17 June 2008 establishing a framework for community action in the field of marine environmental policy. Official Journal of the European Union L, 2008, 164, 19-40.

[7] Napper, I.E., Bakir, A., Rowland, S.J., Thompson, R.C.: Characterisation, quantity and sorptive properties of microplastics extracted from cosmetics. Marine Pollution Bulletin, 2015, 99(1), 178-185.

[8] Gouin, T., Roche, N., Lohmann, R., Hodges, G.: A thermodynamic approach for assessing the environmental exposure of chemicals absorbed to microplastic. Environmental Science & Technology, 2011, 45(4), 1466-1472.

[9] Cheung, P.K., Fok, L.: Evidence of microbeads from personal care product contaminating the sea. Marine Pollution Bulletin, 2016, 109(1), 582-585.

[10] Bergmann, M., Gutow, L., Klages, M.: Marine anthropogenic litter. Springer, 2015.

[11] Kershaw, P.: Sources, fate and effects of microplastics in the marine environment: A global assessment. Reports and studies-IMO/FAO/Unesco-IOC/WMO/IAEA/UN/UNEP Joint Group of Experts on the Scientific Aspects of Marine Environmental Protection (GESAMP), 2015, 90(10.13140).

[12] Andrady, A.L.: Microplastics in the marine environment. Marine Pollution Bulletin, 2011, 62(8), 1596-1605.

[13] Andrady, A., Pegram, J., Searle, N.: Wavelength sensitivity of enhanced photodegradable polyethylenes, ECO, and LDPE/MX. Journal of Applied Polymer Science, 1996, 62(9), 1457-1463.

[14] Astrom, L.: Shedding of synthetic microfibers from textiles. University of Gothenburg, 2016, 378.

[15] Essel, R., Engel, R., Carus, M., Ahrens, R.: Sources of microplastics relevant to marine protection in Germany. Umweltbundesamt Texte, 2015, 64, 2015.

[16] Browne, M.A., Crump, P., Niven, S.J., Teuten, E., Tonkin, A., Galloway, T., Thompson, R.: Accumulation of microplastic on shorelines woldwide: sources and sinks. Environmental Science & Technology, 2011, 45(21), 9175-9179.

[17] Thompson, R.C., Olsen, Y., Mitchell, R.P., Davis, A., Rowland, S.J., John, A.W., McGonigle, D.Russell, A.E.: Lost at sea: where is all the plastic? Science, 2004, 304(5672), 838-838.

[18] Napper, I.E. Thompson, R.C.: Release of synthetic microplastic plastic fibres from domestic washing machines: effects of fabric type and washing conditions. Marine Pollution Bulletin, 2016, 112(1-2), 39-45.

[19] Derraik, J.G.: The pollution of the marine environment by plastic debris: a review. Marine Pollution Bulletin, 2002, 44(9), 842-852.

[20] Browne, M.A., Galloway, T., Thompson, R.: Microplastic — An emerging contaminant of potential concern? Integrated Environmental Assessment and Management, 2007, 3(4), 559-561.

[21] Thompson, R.C.: Plastic debris in the marine environment: consequences and solutions. Marine Nature Conservation in Europe, 2006, 193, 107-115.

[22] Solomon, O.O., Palanisami, T.: Microplastics in the marine environment: current status, assessment methodologies, impacts and solutions. Journal of Pollution Effects & Control, 2016, 1-13.

[23] Browne, M.A., Galloway, T.S., Thompson, R.C.: Spatial patterns of plastic debris along estuarine shorelines. Environmental Science & Technology, 2010, 44(9), 3404-3409.

[24] Moore, C.J., Moore, S.L., Weisberg, S.B., Lattin, G.L., Zellers, A.F.: A comparison of neustonic plastic and zooplankton abundance in southern California's coastal waters. Marine Pollution Bulletin, 2002, 44(10), 1035-1038.

[25] Barnes, D.K., Galgani, F., Thompson, R.C., Barlaz, M.: Accumulation and fragmentation of plastic debris in global environments. Philosophical Transactions of the Royal Society of London B: Biological Sciences, 2009, 364(1526), 1985-1998.

[26] Thompson, R., Moore, C., Andrady, A., Gregory, M., Takada, H., Weisberg, S.: New directions in plastic debris. Science, 2005, 310(5751), 1117-1117.

[27] Nor, N.H.M. Obbard, J.P.: Microplastics in Singapore's coastal mangrove ecosystems. Marine Pollution Bulletin, 2014, 79(1), 278-283.

[28] Gewert, B., Ogonowski, M., Barth, A., MacLeod, M.: Abundance and composition of near surface microplastics and plastic debris in the Stockholm Archipelago, Baltic Sea. Marine Pollution Bulletin, 2017, 120(1-2), 292-302.

[29] Gündoğdu, S., Çevik, C.: Micro-and mesoplastics in Northeast Levantine Coast of Turkey: The preliminary results from surface samples. Marine Pollution Bulletin, 2017, 118(1), 341-347.

[30] Zobkov, M., Esiukova, E.: Microplastics in Baltic bottom sediments: Quantification procedures and first results. Marine Pollution Bulletin, 2017, 114(2), 724-732.

[31] Dekiff, J.H., Remy, D., Klasmeier, J., Fries, E.: Occurrence and spatial distribution of microplastics in sediments from Norderney. Environmental Pollution, 2014, 186, 248-256.

[32] Laglbauer, B.J., Franco-Santos, R.M., Andreu-Cazenave, M., Brunelli, L., Papadatou, M., Palatinus, A., Grego, M., Deprez, T.: Macrodebris and microplastics from beaches in Slovenia. Marine Pollution Bulletin, 2014, 89(1), 356-366.

[33] Thompson, R.C., Moore, C.J., Vom Saal, F.S., Swan, S.H.: Plastics, the environment and human health: current consensus and future trends. Philosophical Transactions of the Royal Society of London B: Biological Sciences, 2009, 364(1526), 2153-2166.

[34] Eriksen, M., Mason, S., Wilson, S., Box, C., Zellers, A., Edwards, W., Farley, H., Amato, S.: Microplastic pollution in the surface waters of the Laurentian Great Lakes. Marine Pollution Bulletin, 2013, 77(1-2), 177-182.

[35] Zylstra, E.: Accumulation of wind-dispersed trash in desert environments. Journal of Arid Environments, 2013, 89, 13-15.

[36] Horton, A.A., Walton, A., Spurgeon, D.J., Lahive, E., Svendsen, C.: Microplastics in freshwater and terrestrial environments: Evaluating the current understanding to identify the knowledge gaps and future research priorities. Science of The Total Environment, 2017, 586, 127-141.

[37] Moore, C., Lattin, G., Zellers, A.: Quantity and type of plastic debris flowing from two urban rivers to coastal waters and beaches of Southern California. Revista de Gestão Costeira Integrada - Journal of Integrated Coastal Zone Management, 2011, 11(1).

[38] McCormick, A., Hoellein, T.J., Mason, S.A., Schluep, J., Kelly, J.J.: Microplastic is an abundant and distinct microbial habitat in an urban river. Environmental Science & Technology, 2014, 48(20), 11863-11871.

[39] Anderson, P.J., Warrack, S., Langen, V., Challis, J.K., Hanson, M.L., Rennie, M.D.: Microplastic contamination in Lake Winnipeg, Canada. Environmental Pollution, 2017, 225, 223-231.

[40] Zbyszewski, M., Corcoran, P.L., Hockin, A.: Comparison of the distribution and degradation of plastic debris along shorelines of the Great Lakes, North America. Journal of Great Lakes Research, 2014, 40(2), 288-299.

[41] Alencastro, D.: Pollution due to plastics and microplastics in Lake Geneva and in the Mediterranean Sea. Journal of Archaeological Science, 2012, 65, 157-164.

[42] Dris, R., Gasperi, J., Rocher, V., Saad, M., Renault, N., Tassin, B.: Microplastic contamination in an urban area: a case study in Greater Paris. Environmental Chemistry, 2015, 12(5), 592-599.

[43] Lechner, A., Keckeis, H., Lumesberger-Loisl, F., Zens, B., Krusch, R., Tritthart, M., Glas, M., Schludermann, E.: The Danube so colourful: a potpourri of plastic litter outnumbers fish larvae in Europe's second largest river. Environmental Pollution, 2014, 188, 177-181.

[44] Klein, S., Worch, E., Knepper, T.P.: Occurrence and spatial distribution of microplastics in river shore sediments of the Rhine-Main area in Germany. Environmental Science & Technology, 2015, 49(10), 6070-6076.

[45] Sadri, S.S. Thompson, R.C.: On the quantity and composition of floating plastic debris entering and leaving the Tamar Estuary, Southwest England. Marine Pollution Bulletin, 2014, 81(1), 55-60.

[46] Free, C.M., Jensen, O.P., Mason, S.A., Eriksen, M., Williamson, N.J.Boldgiv, B.: High-levels of microplastic pollution in a large, remote, mountain lake. Marine Pollution Bulletin, 2014, 85(1), 156-163.

[47] Wang, J., Peng, J., Tan, Z., Gao, Y., Zhan, Z., Chen, Q., Cai, L.: Microplastics in the surface sediments from the Beijiang River littoral zone: Composition, abundance, surface textures and interaction with heavy metals. Chemosphere, 2017, 171, 248-258.

[48] Su, L., Xue, Y., Li, L., Yang, D., Kolandhasamy, P., Li, D., Shi, H.: Microplastics in Taihu Lake, China. Environmental Pollution, 2016, 216, 711-719.

RECENT ADVANCES IN MICROPLASTIC RESEARCH IN JAPAN

Atsuhiko Isobe

Research Institute for Applied Mechanics, Kyushu University, 6-1 Kasuga-koen, Kasuga, 8168580, Japan, aisobe@riam.kyushu-u.ac.jp

Keywords: Pelagic microplastics, standardization/harmonization, field surveys and prediction

Abstract

Recent advances of microplastic (MP) studies in Japan are reviewed. To standardize and to harmonize the oceanic MP measurement, the Ministry of Environment, Japan recently opened a guideline publicly on their website. In line with the protocols in the guideline, the situation was uncovered that the East Asian seas are located in a hotspot of marine plastic pollution, and the situation will continue in the future unless the mismanaged plastic waste is drastically reduced around these regions.

1 Introduction

Recent advances of microplastic (MP) studies in Japan are reviewed. The research projects regarding marine plastic debris started in 2007 under a support by the Environmental Research and Technology Development Fund of the Ministry of the Environment, Japan. A research project supported by the same grant is now on-going as the "Comprehensive studies on oceanic transport, environmental risk, and advanced monitoring of marine plastic debris", which includes monitoring, risk assessment, and numerical modeling of ocean plastic debris especially for MP. In the present paper, monitoring and modeling studies on ocean pelagic MP are summarized.

2 Standardized/harmonized protocols for MP measurement

Field surveys, laboratory analyses, and data processing for MP abundance should be standardized among researchers over the world; otherwise comparisons and synthesis of these data are difficult. Recently, a guideline for MP monitoring was opened publicly on the website of the Ministry of Environment, Japan [1] (Figure 1). The essence of the guideline is briefly described below.

2.1 Field surveys

A neuston (or manta) net, originally designed for sampling zooplankton, fish larvae, and fish eggs near the sea surface, is used for sampling the small plastic fragments. The mouth, length, and mesh size of the net are approximately 75 × 75 cm, 3 m, and 0.35 mm, respectively. The lower limit of the MP discussed in recent studies is, therefore, dependent on this mesh size. A research vessel or boat tows the neuston net around stations continuously for 10-30 min (depending on abundance of suspended sediments) at a constant speed of 2–3 knots. It is desirable to conduct sampling at the side of the vessel with less influence from turbulence induced by vessel motion, and from debris generated from the vessel (Figure 1). A minimum requirement is a flow meter installed around the net mouth (Figure 2) to measure the water volume passing through during the sampling. Otherwise, the estimates of the concentration of small plastic fragments would become inaccurate, because the speed of ocean currents frequently exceeded O (1) knot. Once the surveys are completed, the flow-meter readings and net mouth dimensions (75 × 75 cm) were used to estimate the volume of water filtered during each tow.

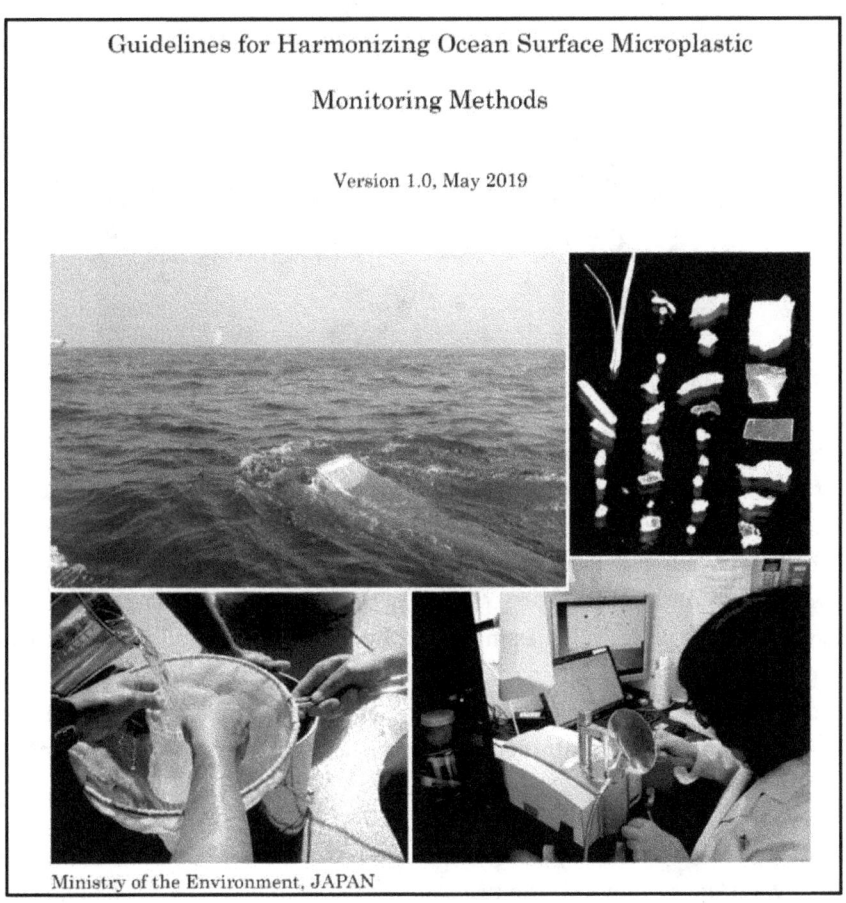

Figure 1: Guideline for pelagic MP measurement opened publicly on the website of the Ministry of Environment, Japan [1]

2.2 Laboratory analyses

A procedure adopted in Kyushu University is briefly described below. The readers are referred to the guideline [1] for a more thorough description of the procedure detail. The small plastic fragments collected are brought back to laboratories to distinguish them from other suspended matter. All samples are first observed on a monitor display via a USB camera (HDCE-20C; AS ONE Corporation) attached to a stereoscopic microscope (SZX7; Olympus Corporation) and identified visually by their color and shape [2]. Polymer types of material were identified using a Fourier transform infrared spectrophotometer (FT-IR alpha; Bruker Optics K.K.), when fragments are too small for visual differentiation between MP and biological matter. The plastic debris are categorized into fragments, filaments, and expanded-polystyrene (hereinafter, EPS) particles.

Figure 2: A flow-meter installed at the mouth of a neuston net

The numbers of plastic pieces (hereinafter "quantity") in each size range are counted with an increment of 0.1 mm for microplastics (MP), 1 mm for mesoplastics between 5 and 10 mm, and 10 mm for mesoplastics >10 mm. The sizes are defined by the longest length of each irregularly shaped plastics visible on the monitor display, as measured using image-processing software (ImageJ downloaded from http://imagej.nih.gov). The quantities within each size range are thereafter divided by the water volumes measured by the flow meter at each sampling station to convert them to quantities per unit seawater volume (hereinafter "concentration" with units of pieces per cubic meter = P/m^3). The concentrations of microplastics and mesoplastics are computed by integrating the concentrations of the plastics with sizes from 0.3 to 5 mm and from 5 to 40 mm, respectively.

An interlaboratory comparison exercise was conducted to assess the consistency of MP quantification across several laboratories [3]. The test samples were prepared by mixing one liter seawater free of plastics, MP made from polypropylene, high- and low-density polyethylene, and artificial particles such as wood chips, bivalve shells, crab shells, eggshells, and cultured zooplankton (*Artemia*) free of plastic fragments in two plastic bottles, and analyzed concurrently in 12 experienced laboratories around the world. The minimum requirements to quantify MP were examined by comparing actual numbers of MP in these sample bottles with numbers measured in each laboratory. The uncertainty was due to pervasive errors derived from inaccuracies in measuring sizes and/or misidentification of MP, including both false recognition and overlooking. The identification and confirmation of plastic materials (i.e., polymer types) using vibrational spectroscopy is especially critical for MP smaller than 2 mm. The number of MP < 1 mm was underestimated by 20% even when using the best practice for measuring MP in laboratories.

2.3 Data processing

The vertical distribution (hence, surface concentration) of light-weighted MP is vulnerable to oceanic turbulence induced by wind and waves. To compare the estimated concentrations of MP obtained under the different oceanic conditions, the concentration (P/m^3) of MP is integrated vertically (P/km^2). The quantities of small plastic fragments decrease exponentially into deeper layers [4, 5], and thus, the vertical distribution of the concentration (N) of MP can be expressed as follows:

$$N = N_0 e^{\frac{w}{A_0} z}, \qquad (1)$$

where N_0 denotes the concentration of MP collected using the neuston net, w is the plastic rise velocity (5.3 mm/s) obtained experimentally by Reisser et al. (2015) [5], and z is the vertical axis looking upward from the sea surface. The parameter A_0 is computed as:

$$A_0 = 1.5\, u_* \, k\, H_s, \qquad (2)$$

where u_* represents the frictional velocity of water (=0.0012 W_{10}), k is the von Karman coefficient (0.4), H_s is the significant wave height, and W_{10} is the 10-m wind speed [4]. Vertically integrating Eq. (1) from the sea surface ($z = 0$) to the infinitely deep layer ($z \rightarrow -\infty$) yields the number of MP per unit area M (P/km^2) as:

$$M = N_0 A_0 / w, \qquad (3)$$

which can be used for comparison with previous studies. We hereinafter refer to M as the "total particle count" in line with Eriksen et al. (2014) [6]. Therefore, it is desirable to concurrently measure wind speed and significant wave height as meta data during the net towing. When these meta data are not obtained during each survey period, satellite wind data such as the Advanced Scatterometer [7] are available. In addition, records of significant wave heights at the nearest observatories are used instead of in-situ data.

3 Field surveys of pelagic MP

3.1 Oceans around Japan

To investigate concentrations of pelagic MP (< 5mm in size) and mesoplastics (> 5mm) in the East Asian seas around Japan, field surveys using five vessels have been conducted since summer 2014. The training vessels belonging to the Tokyo University of Marine Science and Technology (T/V Umitaka-maru and T/V Shinyo-maru), Hokkaido University (T/V Oshoro-maru), Nagasaki University (T/V Nagasaki-maru), and Kagoshima University (T/V Kagoshima-maru) have participated this field campaign supported by the Ministry of Environment, Japan. The MP collected by these vessels are sent to the Kyushu University to measure the MP abundance at each station. The total particle count (P/km^2) is computed based on observed concentrations (P/m^3) of small plastic fragments (both MP and mesoplastics) collected using neuston nets.

The field samplings of MP and mesoplastic have been conducted at more than 300 stations from 2014 to the present (Figure 3). The concentrations averaged over the all stations were 2.5, 0.07, and 0.45 P/m³ for fragments, filaments, and EPS microplastics, respectively. The MP in the fragment type were most abundant and were found especially around Japan (Figure 4). The total particle count of MP was estimated using the concentrations averaged over 2014 data, and was 1,720,000 P/km², 16 times greater than in the North Pacific and 27 times greater than in the world oceans [8].

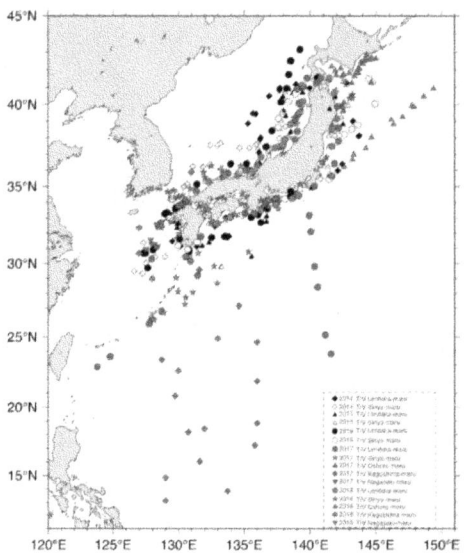

Figure 3: Observation stations for MP survey from 2014 to 2018 using five training vessels in Japan

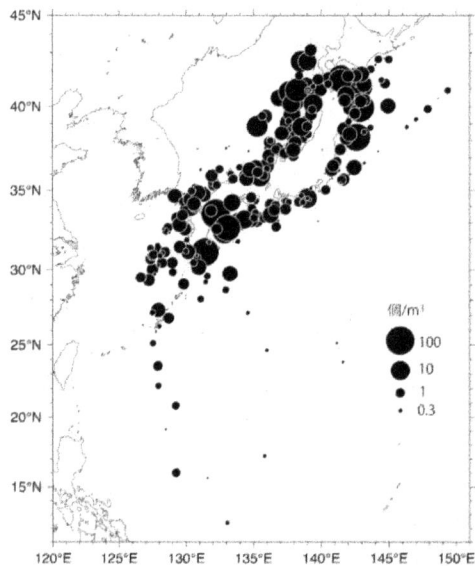

Figure 4: Map of surface concentrations of fragment MP collected in 2014-2018.
The density is proportional to diameter of circles (see references in the lower right corner).

3.2 Pacific Ocean

A transoceanic MP survey was conducted to uncover the meridional variation of pelagic MP abundance in February 2016 (Figure 5a) [9]. The abundance was found to decrease exponentially from the North Pacific, via the equatorial Pacific and Tasman Sea, to the Southern Ocean (Figure 5b and 5c). In total, the number of MP with diameters < 5 mm (mesoplastics > 5 mm) excluding fibers and expanded polystyrene from Sta. 20 to Sta. 38 was 932 (109), accounting for 82% (91%) of those collected across the entire meridional transect. Particle counts per unit seawater volume (surface concentration) decreased, as one moved southward (Figure 5b). An exponential curve ($C_0 10^{\lambda \varphi}$; $-63.5° < \varphi < 34.2°$ in latitude) was fitted to the concentration in a least square sense, where C_0 and λ (slope) were 0.02 P/m^3 and 0.0072 degree^{-1}, respectively, with a significant correlation coefficient (0.42) suggested by the *t* test with a 95% confidence level. Particle counts integrated over the water column (total particle count) were less dispersed than the concentrations because of the wind/wave correction (Figure 5c). An exponential curve, fitted to the total particle count of MP, had C_0 and λ of 7.6x10^4 P/km^2 and 0.0097 degree^{-1}, respectively, with a correlation coefficient of 0.53, which was significant at the 99% confidence level according to a *t* test. Averaging the latter exponential curve over the latitudes of 30° to 40° yielded an approximate total particle count in the mid-latitudes of the South Pacific of around 35,000 P/km^2, an order of magnitude smaller than that in the North Pacific (~160,000 P/km^2). It should be noted that the total particle count at Sta. 38, the northernmost station, reached 8,800,000 P/km^2, which was an order of magnitude larger than in the North Pacific. This was consistent with Isobe et al. (2015) [8], who found that the East Asian seas are a hotspot of pelagic MP.

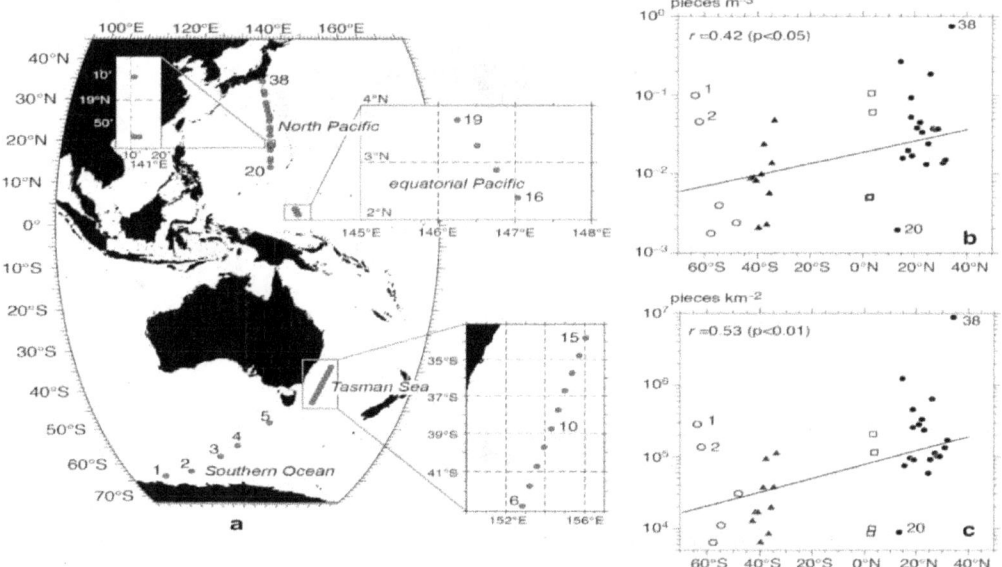

Figure 5: Results of transoceanic survey from the Southern Ocean to Japan in 2016
(After Isobe et al., 2019 [9])

The survey stations along a meridional transect from Sta.1 to Sta. 38 are shown by the dots (a). Areas with a dense network of stations are enlarged in the inset maps. Fig. 5b shows concentrations in the Southern Ocean (open circles), Tasman Sea (closed triangles), Equatorial Pacific (open squares), and North Pacific (closed circles), while Fig. 5c is the same but for total particle counts. The digits 1, 2, 20, and 38 by the marks denote the station numbers. The regression lines between locations versus MP abundance on a $\log_{10}$-scale are shown in each panel with the correlation coefficient (r) and confidence level (e.g., p<0.05 means 95%) suggested by a t test.

3.4 Southern Ocean

Of particular interest in Figure 5b and 5c is that pelagic MP were found even in the Southern Ocean, the most distant ocean from the everyday life generating plastic debris [10]. Five net-tows were performed and 44 pieces of plastic collected. Total particle counts of the entire water column, which is free of vertical mixing, were computed using the surface concentration of MP, wind speed, and significant wave height during the observation period. Total particle counts at two stations near Antarctica were estimated to be in the order of 100,000 P/km^2 (Figure 6). Significant concentration of MP in the Southern Ocean would suggest that marine plastic pollution has spread across the world's oceans. The current surveys revealed a relatively dense concentration of MP in the Southern Ocean, comparable with concentrations observed in the Northern Hemisphere oceans (two of five stations in the present case). The present findings raise concern about the widespread nature of marine plastic pollution, indicating that plastic-free ocean environments are increasingly rare.

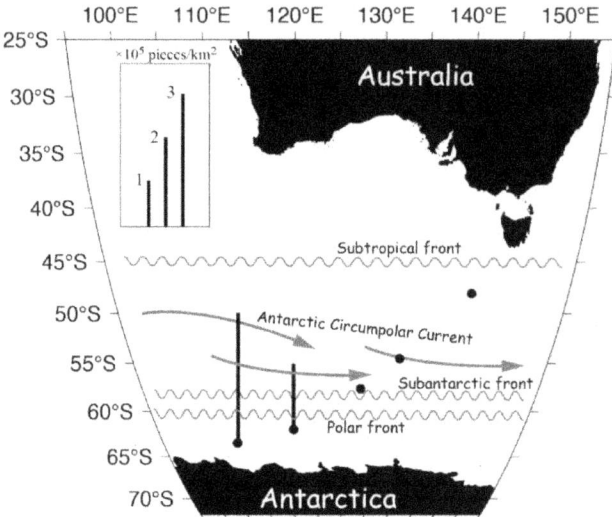

Figure 6: Observed total particle counts and schematic view of oceanic conditions in the Southern Ocean. The bars at Stas. 1 and 2 indicate the total particle counts (see the reference scale shown in the upper left corner). The particle counts at Stas. 3, 4, and 5 were not shown, since the numbers of collected MP were negligibly small. (After Isobe et al., 2017 [10])

4 Future prediction by a numerical simulation

Laboratory-based studies have suggested that marine organisms can be harmed by ingesting MP. However, unless the current and future MP abundance in the ocean environment is quantified, these experimental studies could be criticized for using an unrealistic density or sparsity of MP. The secular variations of pelagic MP abundance in the Pacific Ocean from 1957 to 2066 were computed, based on a combination of numerical modeling and transoceanic surveys conducted meridionally from Antarctica to Japan [9].

Four models were combined to determine the MP distribution over the Pacific Ocean from 60°S to 60°N. The target was MP larger than 0.3 mm that one could observe in the present survey, and that one could use to validate model accuracy. First, an emission model was constructed to generate MP from different sources. Second, surface ocean currents were given by oceanic analysis/reanalysis model products. Third, a wave model was used to compute Stokes drift, which also contributes to MP displacements in conjunction with surface ocean currents [9]. Fourth, a particle-tracking model incorporating a sink term was used to determine the motion of particles representing non-conservative MP drifting in the upper ocean. The emission of the modeled particles at the present time was proportional to the mismanaged plastic wastes in countries surrounding the Pacific Ocean and had a temporal variation proportional to GDP in each region. Also, the future emission was proportional to the 15-year prediction of the plastic wastes provided by Jambeck et al. (2015) [11]. In the present model, the modeled particle count was converted to the total particle count of MP over the model domain. The total particle count was adjusted observed at Sta. 38 (Figure 5a), where MP abundance was greatest along the meridional transect in the 2016 survey, to the modeled particle count in the nearest grid cell in February 2016.

Marine plastic pollution is an ongoing concern especially in the North Pacific, and pelagic MP are regarded as non-conservative matter due to the removal processes that operate in the upper ocean. The results of our numerical model incorporating removal processes on a 3-year timescale suggested that the weight concentration was projected to exceed 1,000 mg/m^3 in summer from the 2060s onward in parts of the East Asian seas and the central North Pacific (Figure 7).

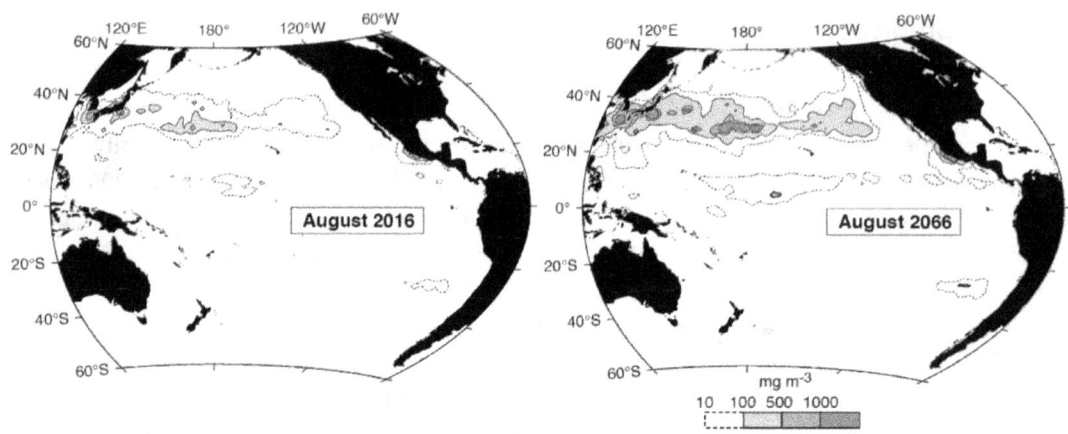

Figure 7: Abundance of MP in the present and future.

The panels represent the weight concentrations averaged for August 2016 (left) and for August 2066 (right) at the sea surface. The weight concentrations are shown by a stippling in the line with the scale at the bottom of the right panel. The broken curves denote a weight concentration of 10 mg/m^3. (After Isobe et al., 2019 [9])

5 Conclusions

The monitoring and modeling of ocean pelagic MP conducted by Japanese researchers are summarized. The recent accomplishment uncovered the situation that the East Asian seas are located in a hotspot of marine plastic pollution, and the situation will continue in the future unless the mismanaged plastic waste is drastically reduced around these regions. The situation is probably the same in the Southeast Asian regions, although the monitoring and modeling efforts have not yet been conducted sufficiently. Recently, a new research project regarding marine plastic pollution in the Southeast Asian seas has started with a cooperation between Japan and Thailand (SATREPS granted by the Japan Science & Technology Agency and the Japan International Cooperation Agency; PI: Isobe), and will hopefully contribute to combat marine plastic pollution in this region.

6 Acknowledgements

Prof. Sandhya Babel and the Exceed Swindon project, who invited the author to the Expert Workshop on microplastics in Koh Samui, Thailand is greatly appreciated. The accomplishments provided in this paper are mostly from the research projects supported by the Environmental Research and Technology Development Fund (4-1502 and SII-2) of the Ministry of the Environment, Japan.

7 References

[1] Michida, Y. Chavanich, S., Cózar A., Hagmann, P., Hinata, H., Isobe, A., Kershaw, P., Kozlovskii, N., Li, D., Lusher, A.L., Martí, E., Mason, S.A., Mu, J., Saito, H., Shim, W. J., Syakti, A.D., Takada, H., Thompson, R., Tokai, T., Uchida, K., Vasilenko, K., Wang, J.: Guidelines for Harmonizing Ocean Surface Microplastic Monitoring Methods. Ministry of the Environment Japan, 2019, 71 pp. (http://www.env.go.jp/en/water/marine_litter/guidelines/guidelines.pdf)

[2] Hidalgo-Ruz, V., Gutow, L., Thompson, R.C., Thiel, M.: Microplastics in the marine environment: a review of the methods used for identification and quantification. Environ. Sci. Technol. 2012, 46, 3060-3075.

[3] Isobe, A., Buenaventura, N.T., Chastain, S., Chavanich, S., Cózar, A., DeLorenzo, M., Hagmann, P., Hinata, H., Kozlovskii, N., Lusher, A.L., Martí, E., Michida, Y., Mu, J., Ohno, M., Potter, G., Ross, P.S., Sagawa, N., Shim, W.J., Song, Y.K., Takada, H., Tokai, T., Torii, T., Uchida, K., Vassillenko, K., Viyakarn, V., Zhang, W.: An interlaboratory comparison exercise for the determination of microplastics in standard sample bottles. Mar. Pollut. Bull. 2019, 146, 831-837

[4] Kukulka, T., Proskurowski, G., Moret-Ferguson, S., Meyer, D.W., Law K.L.: The effect of wind mixing on the vertical distribution of buoyant plastic debris. Geophys. Res. Lett. 2012, 39, L07601.

[5] Reisser, J., Slat, B., Noble K., du Plessis, K., Epp, M., Proietti M., de Sonneville, J., Becker T., Pattiaratchi C.: The vertical distribution of buoyant plastics at sea: an observational study in the North Atlantic Gyre. Biogeosciences 2015, 12, 1249-1256.

[6] Eriksen, M., Lebreton, L.C.M., Carson, H.S., Thiel, M., Moore, C.J., Borerro, J.C., Galgani, F., Ryan, P.G., Reisser, J.: Plastic pollution in the world's oceans: More than 5 trillion plastic pieces weighing over 250,000 tons afloat at sea. PLoS ONE 2014, 9(12): e111913.

[7] Kako, S., Isobe, A., Kubota M.: High-resolution ASCAT wind vector dataset gridded by applying an optimum interpolation method in the global ocean. J. Geophys. Res. Atmos. 2011, 116, D23107.

[8] Isobe, A. Uchida, K., Tokai, T., Iwasaki, S: East Asian seas: a hot spot of pelagic microplastics. Mar. Pollut. Bull. 2015, 101, 618-623.

[9] Isobe. A., Iwasaki, S., Uchida, K., Tokai, T.: Abundance of non-conservative microplastics in the upper ocean from 1957 to 2066, Nat. Comm. 2019, 10, 417.

[10] Isobe, A., Uchiyama-Matsumoto, K., Uchida, K., Tokai, T.: Microplastics in the Southern Ocean. Mar. Pollut. Bul. 2017, 114, 623-626.

[11] Jambeck, J.R., Geyer, R., Wilcox, C., Siegler, T.R., Perryman, M., Andrady, A., Narayan R., Law, K.L.: Plastic waste inputs from land into the ocean. Science 2015, 347, 768-771

PRESENCE AND DISTRIBUTION OF MICROPLASTICS FROM UPSTREAM TO DOWNSTREAM OF CITARUM RIVER, WEST JAVA, INDONESIA

Emenda Sembiring[1], Firdha Cahya Alam[1], Veinardi Suendo[2], Muhammad Reza[2]

[1]*Environmental Engineering Department, Institut Teknologi Bandung, Jalan Ganesha 10, Bandung, Indonesia, 40132, emenda@ftsl.itb.ac.id*

[2]*Chemistry Department, Institut Teknologi Bandung, Jalan Ganesha 10, Bandung, Indonesia, 40132*

Keywords: *Citarum* River, microplastics, occurrence, sediment, upstream downstream, water

Abstract

Occurrence of microplastics (MPs) along *Citarum* River, Indonesia, one of the most polluted rivers in the world, has not been established yet. A grab sampling was conducted to find the occurrence and abundance of MPs in water and sediments of *Citarum* River from upstream (pristine area) to downstream (bay area). The average concentration (± standard deviation) of MPs was 9.5 ± 3.03 particles/L in surface water, and the average concentration of MPs in sediment samples was 3.68 ± 1.5 particles/100 g of dried sediment. Those MPs concentration in surface water of *Citarum* River is significantly higher at the bay than at the upper sites. The most prevailing type of MPs is polyethylene terephthalate (PET), which is known as polyester.

1 Introduction

Microplastics (MPs) are plastic particles with diameter sizes less than 5 mm [1]. MPs can be divided into two types, primary and secondary MPs. Primary MPs have small sizes from their source, such as microbeads in facial cleanser [2]. Secondary MPs are fragmented plastics from larger plastics [3]. Due to the accumulation of MPs from various sources from inland to the ocean, many researches have tried to discover the distribution of MPs in the ocean, estuary, and also freshwater environment [4-6]. Some researchers even already modeled the fate of MPs from upstream to downstream [7]. However, research on MPs has just emerged in Indonesia. According to [8], a MPs study on the deep water sediments in the eastern Indian Ocean nearby Western Sumatera showed that the concentration of MPs increases as closer to the main island and in proportion to the increase of anthropogenic activities. In addition, [9] conducted a MPs distribution study on sediments in *Muara Badak* estuary, in *Kutai Kertanegara* District. This study showed that fragments, films, and fibers are types of dominant MPs in Muara Badak sediments.

Citarum River is the largest and longest river in West Java, Indonesia, which is located from Mount *Wayang*, West Java to *Muara Gembong*, Bekasi. This river has a length of about 297 km, and the area of *Citarum* watershed is around 6,540 km^2. There are three large dams namely *Saguling* Reservoir, *Jatiluhur* Reservoir and *Cirata* Reservoir. *Citarum* River was also known as one of the most polluted rivers in the world, making it interesting to discover the MPs abundance in river water.

2 Materials and Methods

2.1 Sampling Period and Location

Sampling was conducted in July 2018. The first sampling was conducted on July 24, 2018, the second sampling on July 25, 2018, and the third one on July 27, 2018.

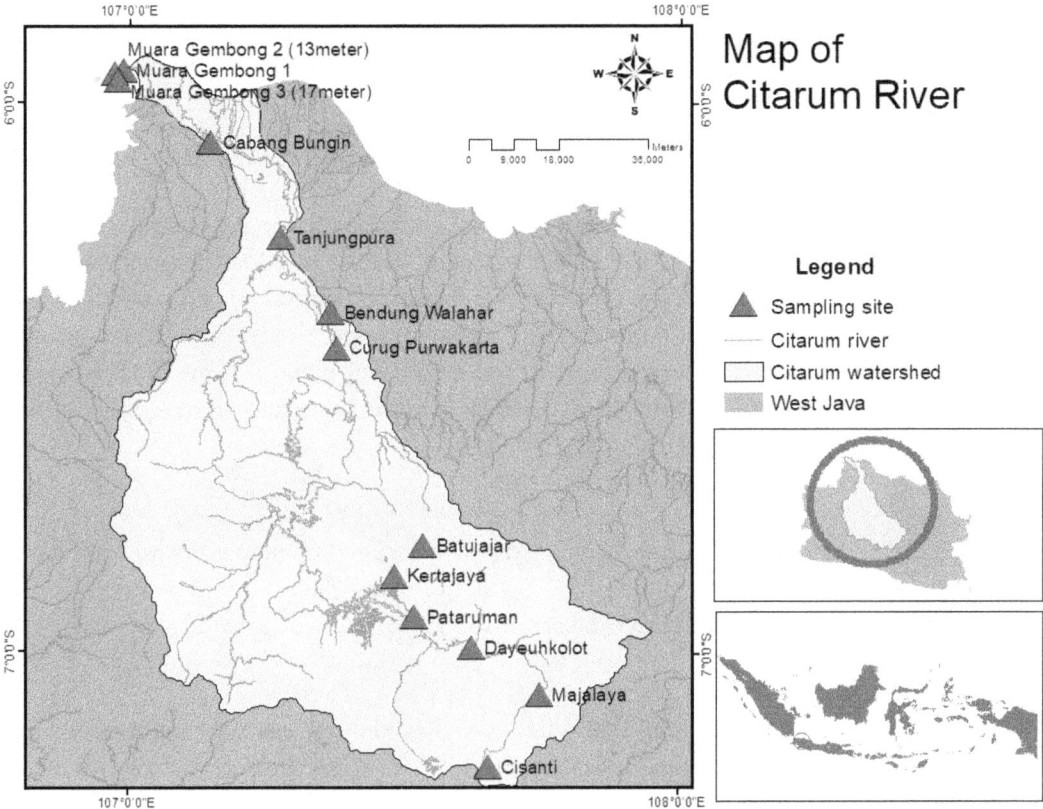

Figure 1: Map of Sampling Locations

These sampling sites were chosen from upstream in *Cisanti, Mount Wayang* until the downstream in the bay of Muara Gembong, Bekasi (Figure 1). Several sampling points were chosen in upstream area, such as in *Majalaya, Dayeuhkolot, Kertajaya, Pataruman,* and *Batujajar,* which are known as densely populated area. Middle part of the river, such as in Curug Purwakarta, Bendung Walahar, and Tanjungpura, with nearby industries and a dam were also selected. Lastly, near the downstream, *Cabang Bungin* and *Muara Gembong*, where mostly fishing activities exist, were chosen.

2.2 Sampling Methods

Sediment and surface water of *Citarum* River were sampled to evaluate the overall distribution of MPs in the river from upstream to downstream and to the sea.

Sediment Sampling

Samples were taken using a shovel for sediments [10]. The sediment samples were then collected in 1 L glass of containers. Sediments samples were taken at the middle part of the river.

Water Sampling

Water samples were collected using grab sampling method by using 1 L glass containers. The containers were first cleansed and rinsed with distilled water. The water samples were taken from the surface of the river [11].

2.3 MPs Separation

Sediment Samples

Sediment samples from each sampling site were processed according to the method reported in [12] with modifications in [13]. First, 1 kg of wet sediment samples was dried in an oven at 100 °C for about 48 h. Then, 100 g of the dried sediment was taken and then suspended with 400 mL 30% NaCl solution. Then, the mixture was stirred using a stirring spoon. Samples were left standing as long as no visible material floated in the supernatant. The floating material in the supernatant was then filtered on Whatman GF/C (glass microfiber filter 1.2 µm) filter paper using a vacuum pump [14].

Water Samples

A subsample of 500 mL was taken from each 1 L sample for duplicate measurement. Each 500 mL water was then filtered using Whatman GF/C (glass microfiber filter 1.2 µm) paper [11]. After filtering, the paper was then dried in an oven at 105 °C for 30 min.

2.4 MPs Visual Inspection

Filter paper was then inspected using a 10-times magnification light binocular microscope brand OPTIKA series B-383FL to find MPs particles. MPs particles were inspected using a microscope and grouped into two types, namely fiber, and fragments. Particles found were counted per 100 g of dry sediment for sediment samples and per 1 L for surface water samples.

2.5 Raman Characterization

Some MPs particles were analyzed by using Raman spectroscopy (Bruker Senterra Raman Spectrometer). A silicon plate was used to place the MPs from the filter paper. To attach them, an ethanol and acetone solution was used. The MPs on the silicon plate was captured with Infinity-X camera and then exposed to light with a wavelength of 785 nm, 25 mW, and integration time 30 s. The observed wave numbers were 100-2500 cm^{-1}. The obtained Raman spectra were then corrected with the background and compared with reference polymer spectra [15-17].

2.5 Data Analysis

Processing and presenting data were conducted with Microsoft Excel 2013 and software R version 3.4.4. One-Way ANOVA was conducted to test, whether there were significant differences in the concentration of MPs particles between sampling points.

3 Results and Discussion

3.1 MPs Concentrations

From the inspection, it was found that the average (± standard deviation) of MPs concentration was 9.5 ± 3.0 particles/L of river water, and the average concentration of MPs in sediment samples was 3.68 ± 1.05 microplastic particles/100 g of dry sediment.

The microplastic concentrations found in the *Citarum* River were higher than those in North Yellow Sea, China, with MPs abundance of 0.54 ± 0.28 particles/L in surface sea water and 3.71 ± 4.27 particles/100 g of dry sediments [18]. In Pearl River, Guangzhou, China, the abundances of MPs in surface water and sediments consecutively ranged from 0.38 to 7.92 particles/L and 8.0 to 959 particles/100 g [19]. If compared with Citarum River, then the number of MPs particles in surface water is lower but in sediments samples is higher.

Citarum River showed less MPs abundance, if compared with Thames River, UK. In Thames, MPs were found in average 66 particles/100 g of dry sediment [20], in Chinese river sediments 80.2 ± 59.4 MPs particles/100 g dry weight [21], and 12.1 ± 0.9 particles/100 g dry sediment on the *Changjiang* Estuary, China [13]. This variation of MPs abundances might be affected by multiple factors such as different methods used or even different river conditions. The different location will show different wave and wind conditions, and light weight MPs can be influenced by vertical mixing caused by oceanic turbulence and the vertical distribution (if observation was conducted using a Neuston net) [22]. Lack of standardized methods makes it hard to directly compare concentration between locations without further reading the methods.

3.2 MPs Visual Inspection on Microscope

The sample visual observation used a 10x lens magnification light microscope (100x total magnification) on Whatman GF/C filter paper (Figure 2). Particles suspected as MPs were indicated from their shape or color that differ from the environment [20].

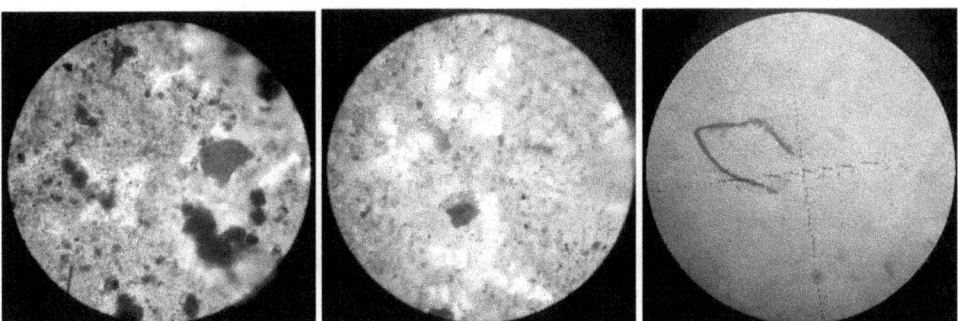

Figure 2: MPs images under light microscope

3.3 MPs Shapes

Based on observations in river water samples, fiber particles were found more (81.3%) than the fragment forms (18.6%). Likewise, for sediment samples, fiber forms were more dominant (86.4%)

than fragment forms (13.5%). These results were similar to [5] and [19] that fibers were the most prevail microplastic shapes in both water and sediment samples. Mostly, if fibers were more abundant, they can be derived from textile wastewater, but higher levels of fragments in surface water suggest additional pathways such as from storm water runoff [23].

3.4 MPs Particles Sizes

The size of the MPs particles observed on inspection using a light microscope is limited from 50 μm to 4,000 μm to avoid false identification of any particle smaller than 50 μm. In Figure 3, one can see the percentage of MPs sizes from water and sediment samples.

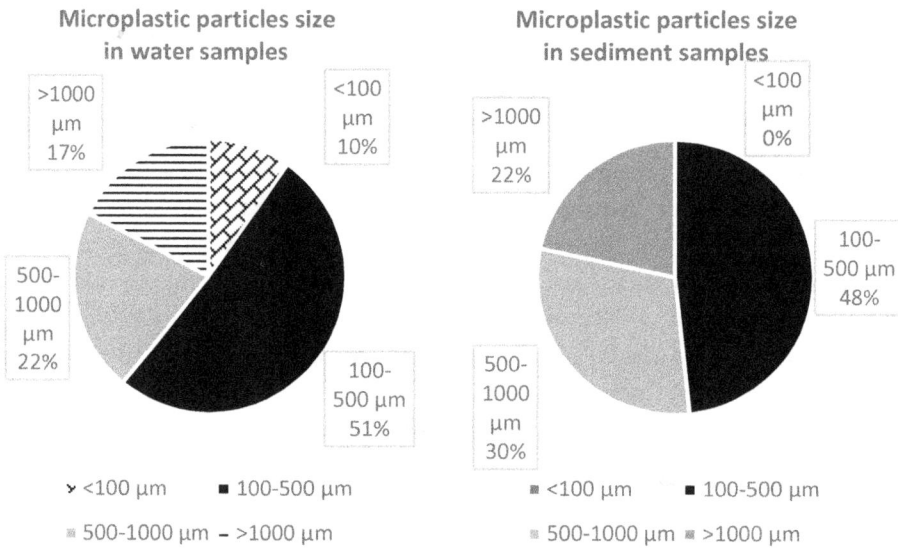

Figure 3: Particles size distribution of MPs in water and sediment samples

It shows that both in the water and sediment samples, MPs particles in the size range of 100-500 μm were found dominant than the other sizes (Figure 3). However, small MPs with sizes less than 100 μm were found only in water samples, but not in sediment samples. It is, because small particles with low density will tend to float on water [24]. Besides, MPs with sizes higher than 1000 μm were found relatively more in sediments than in water samples. Sediment of river with low stream can be a hotspot for deposition of MPs, because larger densities tend to retain in sediment [25]. Generally, settling of MPs can be influenced by their density and morphology [26].

3.5 MPs Concentration at Different Sampling Sites

Concentrations of MPs from upstream to downstream were varying (Figure 4).

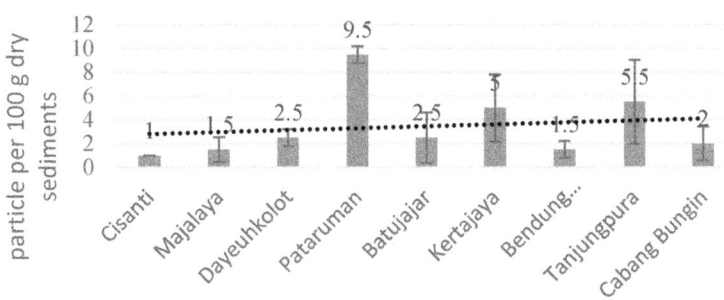

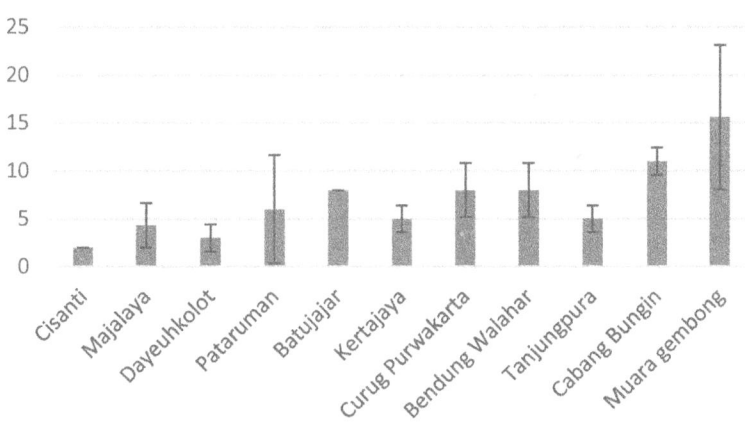

Figure 4: Comparison of the amount of MPs from all sampling sites

It can be seen that MPs concentration were increasing from upstream to downstream and significantly higher in the downstream (p-value = 4.56×10^{-14} < 0.05). This indicated that MPs are floating through the river and accumulating in the bay. Similarly, MPs concentration in sediment samples showed significant differences among locations (p-value = 0.001 < 0.05). It can be seen that concentration in *Pataruman* was relatively higher than at the other locations. One of the reasons can be attributed to the higher industrial activity in *Pataruman*, and also to the domestic dumpsite near the river. The high variability of MPs concentrations indicates that the MPs distribution were influenced by several factors, such as river flow characteristics, distance to source(s), buoyancy behavior, degradation, etc. [27].

In wastewater treatment plants (WWTP), most MPs are removed, retained in the solid residue that may retain in terrestrial ecosystems, if used as fertilizer, and some still remain in the final effluent entering the aquatic ecosystems [28]. However, mostly the industries around upstream *Citarum* have no proper facilities of WWTP. So, the effluent discharge from industry can be one of the sources of MPs in the *Citarum* River. Moreover, MPs pollution can also come from diffuse sources, such as plastics entering a water body through surface run-off, rainfall or wind [26]. Other than that, poor sanitation conditions also tend to make waste dumped directly to the *Citarum* River.

3.5 Raman Identification

The results of visual identification were then tested using Raman spectroscopy. One of observed fibers are shown in Figure 5. Red fiber indicates polyester type as shown from peaks at 631 cm^{-1}, 857 cm^{-1}, 1275 cm^{-1}, 1614 cm^{-1}, and 1727 cm^{-1} [16].

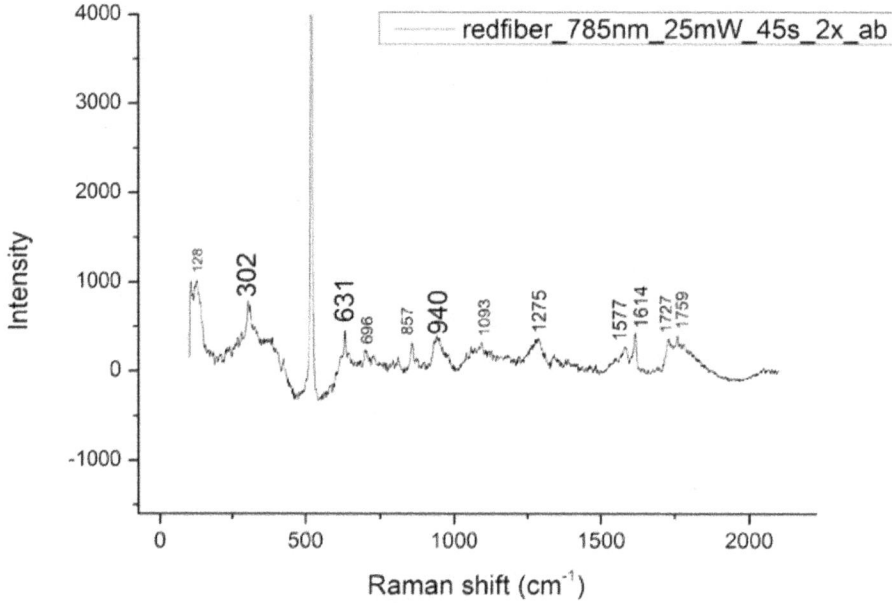

Figure 5: Raman spectra results of red fiber

Polyester is also known as polyethylene terephthalate. This polymer type is common MPs, which is found in wastewater effluents from textile industry [29]. This type of polyester can be distributed to Arctic Ocean [30]. Beside the degradation from clothing, partially UV-degraded post-consumer plastics will also contribute to the different MPs in environment [31].

4 Conclusions

This research showed that MPs concentration in surface water were increasing from upstream to downstream, but did not show a clear pathway in sediments. It was found that the average (± standard deviation) MPs concentrations were 9.5 ± 3.03 particles/L river water and 3.68 ± 1.05 particles/100 g dry sediment. Those MPs concentrations in surface water of *Citarum* River were significantly higher in the bay than at other sites. In addition, MPs concentrations in sediment were found to be higher near industrial area and domestic landfill site.

5 Acknowledgements

This research is funded by Program of Research, Community Service, and Innovation Institute of Technology Bandung (P3MI-ITB). The authors would like to thank DAAD and the project Exceed Swindon to support the participation at the expert workshop on Microplastics in Thailand.

6 References

[1] Arthur, C., Baker, J., Bamford, H.: Proceedings of the International Research Workshop on the Occurrence, Effects, and Fate of Microplastic. Marine Debris, 2009, Group 530.

[2] Fendall, L.S., Sewell, M.A.: Contributing to marine pollution by washing your face: Microplastics in facial cleansers. Marine Pollution Bulletin 2009, 58, 1225–1228.

[3] Browne, M.A., Galloway, T., Thompson, R., Chapman, P.M.: Microplastic an emerging contaminant of concern? Integrated Environmental Assessment and Management 2007, 3, 559-566.

[4] Alves, V.E.N., Figueiredo, G.M.: Microplastics in the sediments of a highly eutrophic tropical estuary. Marine Pollution Bulletin 2019, 58, 1225–1228.

[5] Miller, R.Z., Watts, A.J.R., Winslow, B.O., Galloway, T.S.: Mountain to the sea: River study of plastic and non-plastic microfiber pollution in the norteast USA. Marine Pollution Bulletin 2017, 124, 245-251.

[6] Rech, S., Macaya-Caquilpán, V., Pantoja, J.F., Rivadeneira, M.M., Jofre Madariaga, D., Thiel, M.: Rivers as a source of marine litter — A study from the SE Pacific. Marine Pollution Bulletin 2014, 82, 66–75.

[7] Siegfried, M., Koelmans, A.A., Besseling, E., Kroeze, C.: Export of microplastics from land to sea. A modelling approach. Water Resource 2017, 127, 249–257.

[8] Cordova, M.R., Wahyudi, A.J.: Microplastic in the Deep-Sea Sediment of Southwestern Sumatran Waters, Marine Resources Indonesia 2016, 41, 27-35.

[9] Sari Dewi, I., Aditya Budiarsa, A., Ramadhan Ritonga, I.: Distribusi mikroplastik pada sedimen di Muara Badak, Kabupaten Kutai Kartanegara. Depik 2015, 4, 121–131.

[10] Smith, J.A., Hodge, J., Garver, J.I.: The Distribution of Microplastic Pollution in the Mohawk River. Proceedings of the Mohawk Watershed Symposium, 67 Union College, Schenectady, NY. 2017, 67-71

[11] Barrows, A.P.W., Neumann, C.A., Berger, M.L., Shaw, S.D.: Grab vs. neuston tow net: A microplastic sampling performance comparison and possible advances in the field. Analytical Methods, 2017, 9, 1446–1453.

[12] Wang, J., Peng, J., Tan, Z., Gao, Y., Zhan, Z., Chen, Q., Cai, L.: Microplastics in the surface sediments from the Beijiang River littoral zone: Composition, abundance, surface textures and interaction with heavy metals. Chemosphere, 2017, 171, 248–258.

[13] Peng, G., Zhu, B., Yang, D., Su, L., Shi, H., Li, D.: Microplastics in sediments of the Changjiang Estuary, China. Environmental Pollution, 2017, 225, 283–290.

[14] Hang, K., Su, J., Xiong, X., Wu, X., Wu, C., Liu, J.: Microplastic pollution of lakeshore sediments from remote lakes in Tibet plateau, China. Environmental Pollution, 2016, 219, 450–455.

[15] Cho, L.-L.: Identification of textil fiber by Raman microspectroscopy. Forensic Sciences Journal, 2007, 1, 55–62.

[16] Hager, E., Farber, C., Kurouski, D.: Forensic identification of urine on cotton and polyester fabric with a hand-held Raman spectrometer. Forensic Chemistry, 2018, 9, 44–49.

[17] Was-Gubala, J., Machnowski, W.: Application of Raman spectroscopy for differentiation among cotton and viscose fibers dyed with several dye classes. Spectroscopic Letters, 2014, 47, 527–535.

[18] Zhu, L., Bai, H., Chen, B., Sun, X., Qu, K., Xia, B.: Microplastic pollution in North Yellow Sea, China: Observations on occurrence, distribution and identification. Science of The Total Environment, 2018. 636, 20–29.

[19] Lin, L., Zuo, L.Z., Peng, J.P., Cai, L.Q., Fok, L., Yan, Y., Li, H.X., Xu, X.R.: Occurrence and distribution of microplastics in an urban river: A case study in the Pearl River along Guangzhou City, China. Science of The Total Environment, 2018, 644, 375–381.

[20] Horton, A.A., Svendsen, C., Williams, R.J., Spurgeon, D.J., Lahive, E.: Large microplastic particles in sediments of tributaries of the River Thames, UK – Abundance, sources and methods for effective quantification. Marine Pollution Bulletin, 2017, 114, 218–226.

[21] Peng, G., Xu, P., Zhu, B., Bai, M., Li, D.: Microplastics in freshwater river sediments in Shanghai, China: A case study of risk assessment in mega-cities. Environmental Pollution, 2018, 234, 448–456.

[22] Isobe, A., Uchiyama-Matsumoto, K., Uchida, K., Tokai, T.: Microplastics in the Southern Ocean. Marine Pollution Bulletin, 2017, 114, 623–626.

[23] Sutton, R., Mason, S.A., Stanek, S.K., Willis-Norton, E., Wren, I.F., Box, C.: Microplastic contamination in the San Francisco Bay, California, USA. Marine Pollution Bulletin, 2016, 109, 230–235.

[24] Di, M., Wang, J.: Microplastics in surface waters and sediments of the Three Gorges Reservoir, China. Science of The Total Environment, 2018, 616–617, 1620–1627.

[25] Nizzetto, L., Bussi, G., Futter, M.N., Butterfield, D., Whitehead, P.G.: A theoretical assessment of microplastic transport in river catchments and their retention by soils and river sediments. Environmental Sciences: Processes & Impacts, 2016, 18, 1050–1059.

[26] Siegfried, M., Koelmans, A.A., Besseling, E., Kroeze, C.: Export of microplastics from land to sea. A modelling approach. Water Resource, 2017, 127, 249–257.

[27] Wang, Z., Su, B., Xu, X., Di, D., Huang, H., Mei, K., Dahlgren, R.A., Zhang, M., Shang, X.: Preferential accumulation of small (< 300 µm) microplastics in the sediments of a coastal plain river network in eastern China. Water Resource, 2018, 144, 393–401.

[28] Prata, J.C.: Microplastics in wastewater: State of the knowledge on sources, fate and solutions. Marine Pollution Bulletin, 2018, 129, 262–265.

[29] Hager, E., Farber, C., Kurouski, D.: Forensic identification of urine on cotton and polyester fabric with a hand-held Raman spectrometer. Forensic Chemistry, 2018, 9, 44–49.

[30] Kanhai, L.D.K., Officer, R., Lyashevska, O., Thompson, R.C., O'Connor, I.: Microplastic abundance, distribution and composition along a latitudinal gradient in the Atlantic Ocean. Marine Pollution Bulletin, 2017, 115, 307–314.

[31] Lenz, R., Enders, K., Stedmon, C.A., MacKenzie, D.M.A., Nielsen, T.G.: A critical assessment of visual identification of marine microplastic using Raman spectroscopy for analysis improvement. Marine Pollution Bulletin, 2015, 100, 82–91.

PROSPECTS OF MICROPLASTICS IN THE SEDIMENTARY DEPOSITS ON PATOS LAGOON COAST

Eduardo Saldanha Vogelmann[1], Vitor Rodrigues Alegrini[1], Gabriel Oladele Awe[2], Juliana Prevedello[3]

[1]Institute of Biological Sciences, Federal University of Rio Grande, Avenida Marechal Floriano, 369, ZIP 96170-000, São Lourenço do Sul, Rio Grande do Sul, Brazil;
eduardovogelmann@furg.br

[2]Department of Soil Resources and Environmental Management Department, State University, Ado Ekiti, Nigeria;

[3]Institute of Oceanography, Federal University of Rio Grande, São Lourenço do Sul, Brazil;

Keywords: Water quality, water pollution, environmental pollution, micoplastics

Abstract

Despite the increasing number of studies addressing environmental problems arising from microplastics (MP) in Brazil, the research about this subject is recent and there are currently no studies evaluating the presence of MP as well as their distribution in the southern country region. The objective of this study was to identify and to quantify MP in the sedimentary deposits at different points of the coastal area of São Lourenço do Sul, southern Brazil. For this, sediment samples were collected in seven points at the border of Patos Lagoon in São Lourenço do Sul. At each sampling point, two sediment samples were collected at depths of 0-2 and 10-12 cm. In the laboratory, the sediment samples were disposed in a saline solution (300 g/L NaCl) to separate the supernatant; hydrogen peroxide was added to the fractions after drying to destroy organic material. After this, the fractions were immersed in ethanol solution, adding successive volumes of a saline solution by varying the solution density and adjusting the resultant mixture density to 0.8, 1.0 and 1.2 g/L. The new supernatant was separated and the fragments number was counted using a stereomicroscope. In most part of city coast, MP are incorporated in the superficial (0-2 cm) or subsurface (10-12 cm) sediment. However, in the region near the mouth of São Lourenço stream had the largest MP amount incorporated into the sediment. The predominance of MP with a density between 0.8 and 1.0 g/cm³ was observed.

1 Introduction

Actually, we are living in a "plastic age" with more than 240 million tons of plastic used annually, the majority of which are for disposable use [1]. According to data from the European Plastics Association [2], the world plastics production increased from 5 million in

1950 to 265 million tons in 2010, representing a steady increase of 6% per year over the last 20 years. Currently, plastics are one of the most abundant residues in urban areas mainly due to its physicochemical characteristics, which give it high durability in the environment [3] and its potential contamination of the trophic chain at different levels [4]. Due to their high durability, these materials remain in the environment for many years [1]. This durability is partially based on plastic being an uncommon target for bacteria, which makes it non-biodegradable. However, the residue becomes slowly degraded over time, mainly by exposure to ultra-violet (UV) radiation (heat), wind, rain and other mechanical forces [3]. This results in the fragmentation of particulate plastics into MP [4].

Studies on the composition of plastics in water and beaches accounted for a wide variety of polymers, the most abundant being polypropylene, polyethylene, polyvinyl chloride and polystyrene [5]. Despite this variability in abundance, the vast majority of the polymers can withstand a wide variety of uses and are not biodegradable, although photodegradation and mechanical abrasion contribute to their division/breakdown into smaller particles [6].

Any measure that makes the plastic ages and fragments, particles can sink and become available to benthic organisms. It is not known, what is the degradation rate or how long the plastic may remain in the oceans [7]. However, there exist records of 20 μm particles [5] found on the beaches, dimensions that are identical to those of food items of many invertebrate species, so plastic particles are potentially ingested by these species [8].

There are several records of plastic fragments found in stomach contents of various species of marine birds and mammals, and fish probably as a result of transmission via the food chain, and their effects are unknown [3, 5]. This is worrisome, because the transfer of plastic particles from the digestive system to the circulatory system can damage important organs such as the heart or liver, in addition to the additional threat posed by the effects of toxic substances adsorbed to the MP [9]. In this context, the debate on the silent pollution by MP has been widely heard lately, although not yet physically seen and evident, but is present in the sea, rivers, beaches, and in the organism of vertebrates and invertebrates [10].

In Brazil, research on MP is recent and there are currently no studies evaluating the presence of MP as well as their distribution in the southern region of the country. Therefore, the objective of this study was to identify and to quantify MP in the sedimentary deposits at different points of the coastal area of São Lourenço do Sul, southern Brazil.

2 Material and Methods

The municipality of São Lourenço do Sul is located in the State of Rio Grande do Sul, covering an area of 2,036,125 km^2, with an estimated population of 44,580 inhabitants [11]. The eastern side of the city is bathed by the Patos Lagoon (Figure 1a and b).

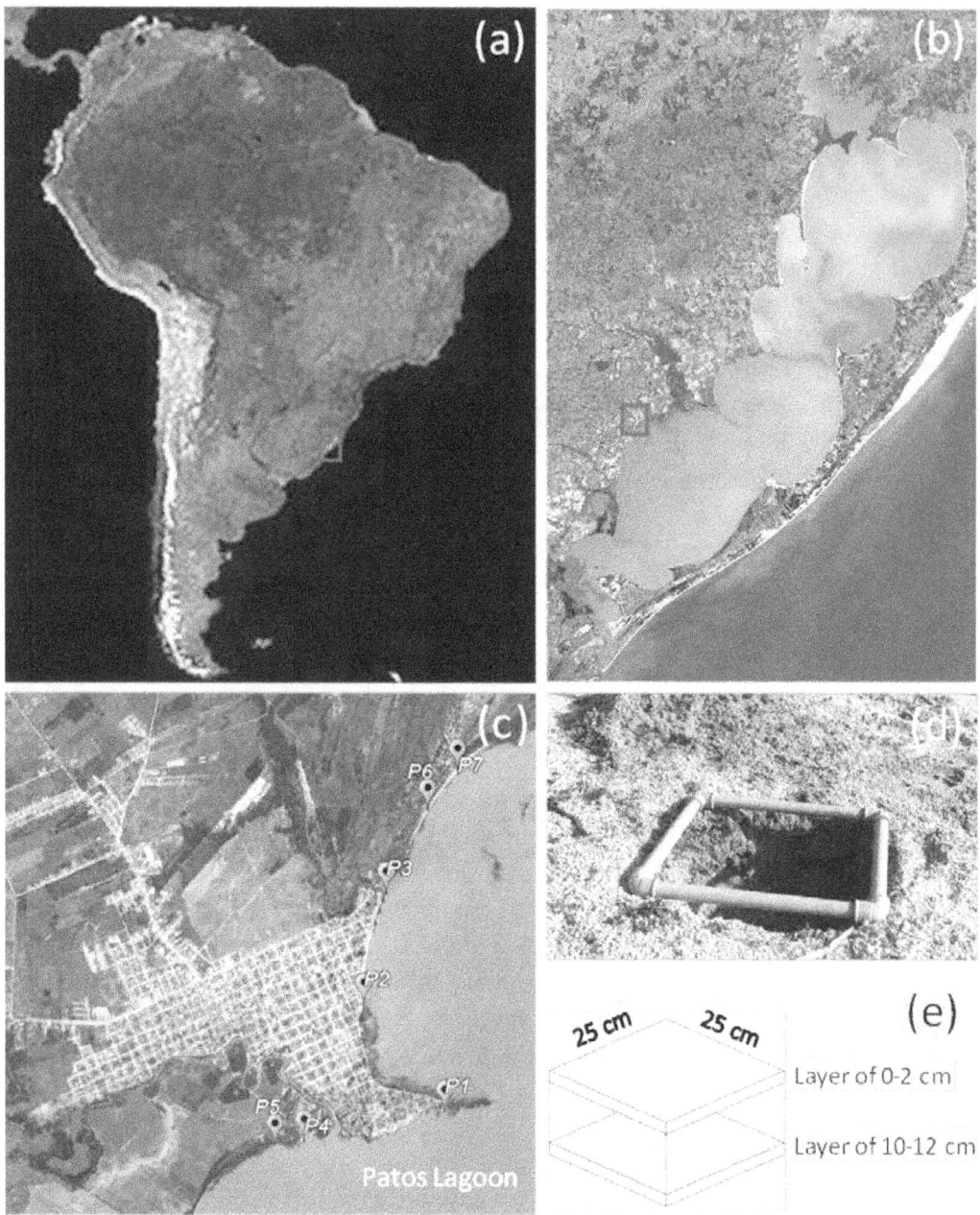

Figure 1: (a) Location of Patos Lagoon in Latin America; (b) Location of São Lourenço do Sul and sampling region in Patos Lagoon; (c) Location of sampling points in the bank of Patos Lagoon; (d, e) Collection scheme of sediment samples in two different layers.

Seven collection points were delineated on the bank of the lagoon (Figure 1c). For MP investigation and quantification, sediment samples were also collected in the region near the limit of the water line (Figure 1c). An area, a square of about 0.25 m^2, was marked in each sampling point for sediment sample collection [12]. At each sampling point, two sediment samples were collected at depths of 0-2 and 10-12 cm (Figure 1d and b).

In the laboratory, the sediment samples were air-dried. After passing a 2 mm mesh, the larger fraction was separated. Then, the < 2 mm sample was disposed in a saline solution (300 g/L NaCl); the mixture was shaken with a glass rod and allowed to stand [12]. After 10 min, the supernatant was separated and air-dried [12, 13]. To the supernatant material, about 25 mL of hydrogen peroxide was applied to destroy organic material [13]. After this, the material was again air-dried and the fragments found were reserved in a vial glass. Subsequently, the fragments were immersed in ethanol solution (96%), adding successive volumes of a saline solution (300 g/L NaCl) by varying the solution density and adjusting the resultant mixture density to 0.8, 1.0 and 1.2 g/L. At each addition of saline solution, the solution was shaken and allowed to rest for 10 min, after this another supernatant was collected. Each fraction of the new supernatant was separated into a vial and the fragments number was counted using a stereomicroscope.

3 Results and Discussion

A great variation was observed between the MP number found between the different points analyzed (Figure 2). However, it is worth to note, a large number of MP found in point 4 was significantly different from the other points. It should be pointed out that this collection point is close to the entry point of São Lourenço stream, a region where the city urbanization was reported to have begun historically. It is the region with the largest population and, consequently, more liable to having more fragments at the river banks [3].

In contrast, at points 6 and 7, smaller fragments volume was observed. However, this is a very sparsely inhabited region, where waste was not expected due to the recent occupation and a low number of residents and tourists, who frequent this area. In points 1 and 3, no MP were observed. This observation may be due to the fact that these regions were recently urbanized, since until the 1990s, practically there were not many tourists visiting these places. Currently, there is a great occupation by tourists especially during the summer, but it is associated with the intense action of a public waste collection of garbage discarded indiscriminately on the banks of the lagoon.

In this regard, it is important to note that MP pollution is affecting large areas and is not concentrated in regions, where there are historically greater anthropic activities or near the site of plastic waste disposal. Recent study by [14] pointed to contamination by MP of different colors and compositions in the Antarctic seas, a practically uninhabited place with no plastic waste disposal points. On this premises, it could be infer that in several parts of the Patos Lagoon, it is possible to find MP in sediments disposed of on the banks and in the bottom sediment of the lake. It is also possible to state that parts of the waste found on the beaches of São Lourenço do Sul may come from waste initially deposited in distant places, given the great ease of plastic waste moving in water bodies, a mobility that is facilitated by the decomposition of smaller fragments that are carried by water, especially the low density MP (polyethylene and polypropylene). This is confirmed in a study conducted by [15], who reported that the dynamics of MP in water are controlled by their physical characteristics (density, size and shape) as well as the dynamic conditions of the oceans (wind, waves, tides, thermo-linear gradients and benthic sediment), which can travel long distances between transport times and when it is deposited on sediment along the banks or bottom of rivers and lakes. Thus, it is evident that the movement of these particles is not clearly defined, because in addition to the intrinsic factors of each MP, these particles can be subjected to stranding, surface drift, vertical mixing and biofouling as well as suspended and suspended load transport processes until they reach terminal deposition on beaches, coastal marshes, benthic sediments, or until they are transported to the oceans [15, 16].

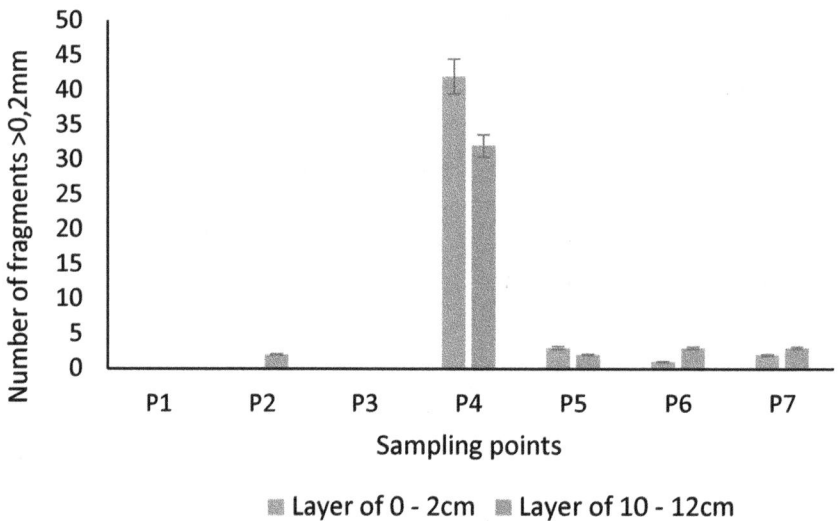

Figure 2: Distribution of plastic fragments smaller than 2 mm in sediment samples

In relation to fragments identified, it is observed that MP predominates with density between 0.8 and 1.0 g/cm³ (over 70%) and lower proportion of MP with density between 1.0 and 1.2 g/cm³. They were observed at all points where MP were found (Figure 3). It is possible that MP predominance in the lighter fraction is associated with low and high-density polypropylene and polyethylene waste, the raw material commonly used in manufacture packing and plastic bags [4]. According to [17], the higher concentration of low density MP is associated with the greater use of polyethylene and polypropylene in confectioneries and products used daily in large quantities by the people, especially in plastic packaging, which in Brazil has no value for the plastic recyclers due to the low value paid for the waste, which is often discarded.

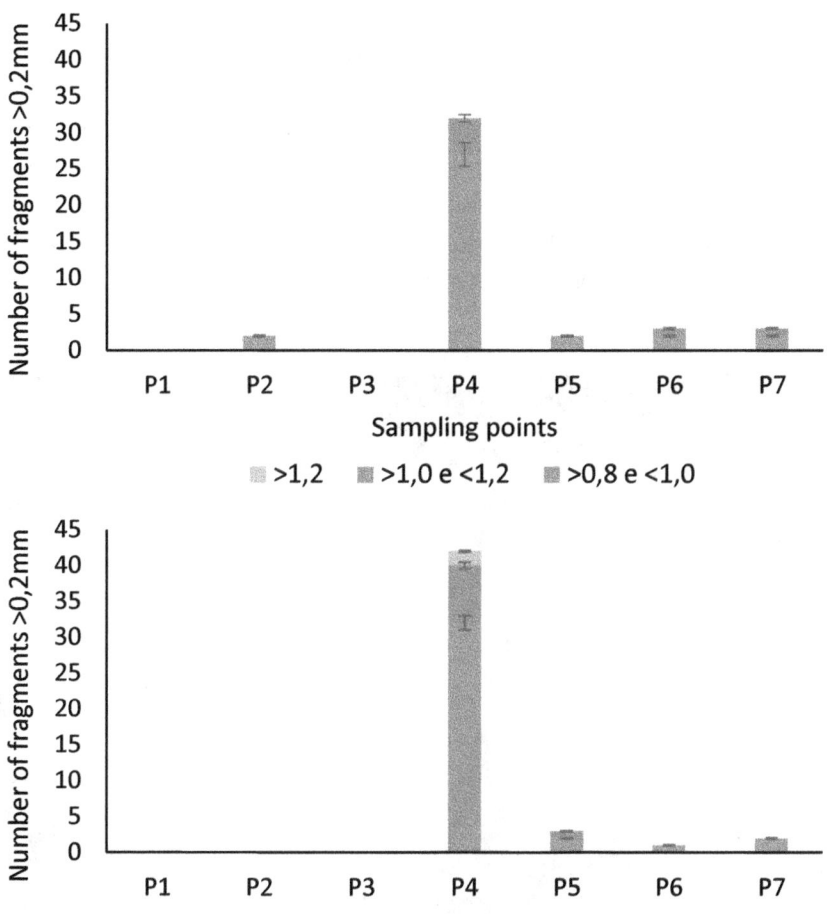

Figure 3: (*top*) Distribution of plastic fragments as function of their density (g/cm³) found in the sediment samples in the 0-2 cm layer, (*below*) Distribution of plastic fragments as function of their density (g/cm³) found in the sediment samples in the 10-12 cm layer

No relationship was observed between the amount of MP fragments and the sampling depths. The predominance of MP fragments in the lower density range (<1.0 g/cm³) may be associated with the fact that MP with high density (>1.2 g/m³) are probably in greater abundance in the bottom sediment of the stream and the Patos Lagoon, thus having their deposition and accumulation in smaller proportion in the sediment deposited on the banks, where smaller MP (<1.0 g/cm³) are concentrated that float and deposit mainly in the coastal region.

Among the main types of plastics observed are plastics of different colors and polymers, which could be attributed to the great variability of residues, from which the micro-residues originate (Figure 4). Probably, all plastics found in all the regions come from inadequate disposal by the people, who also consume products with the most diverse plastic components, and this great variability marks the impact of the urban population. Among the main components, it was possible to observe mainly polyethylene and polypropylene. High density plastics were not found at high concentration, because the methodology used limits the maximum density to 1.2 g/L. However, it can be said that they are not present in the environment, although it is expected that, as they have high density, they are more abundantly deposited in the bottom sediment of the river and the lagoon. In line with this, one can state that sediments are a useful record of past and present plastic leakage from the waste stream into the marine environment [18].

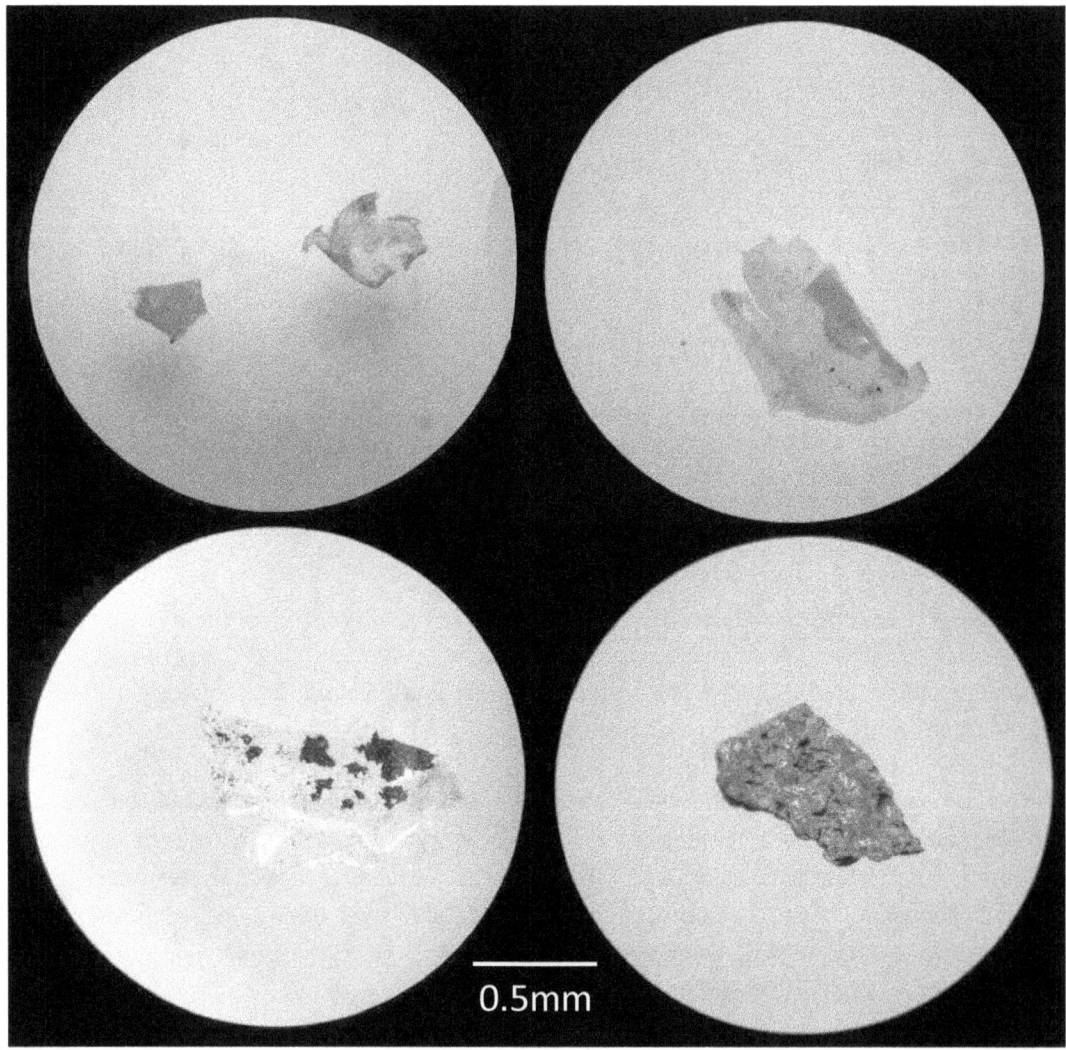

Figure 4: Principal types and colors of the MP observed in the samples

4 Conclusions

Based on the results, it is possible to conclude that in most part of city coast there are MP incorporated in the superficial (0-2 cm) or subsurface (10-12 cm) sediment, dating back to deposits over the years. The region near the point of entry of São Lourenço stream had the largest MP amount incorporated into the sediment, which must come from plastic waste loaded by the stream, since at the collection site there is no intense occupation compared to other sites that presented fewer fragments. It was observed that MP predominated with a density between 0.8 and 1.0 g/cm^3, while other fragments of light fraction polypropylene and polyethylene are of low or high density.

5 Acknowledgements

The authors thank the German Federal Ministry for Economic Cooperation and Development (BMZ), German Academic Exchange Service (DAAD) and the Exceed SWINDON Project for the financial support. This study was also funded by FAPERGS through by the Institutional Scientific Initiation Scholarship Program the PROBIC/FAPERGS 2017-2018.

6 References

[1] Thompson, R.C., Swan, S.H. Moore, C.J., vom Saal, F.S.: Our plastic age. Philosophical Transactions of the Royal Society B 2009, 364, 1973–1976.

[2] Plastics Europe Association: The compelling facts about plastics 2011 - An analysis of plastics production, demand and recovery for 2010. Plastics Europe - Association of Plastics Manufacturers, Brussels, Belgium, 2010, 24 p.

[3] Derraik, J.G.B.: The pollution of the marine environment by plastic debris: a review. Marine Pollution Bulletin 2002, 44, 842–852.

[4] Morét-Ferguson, S.E., Law, K.L., Proskurowski, G., Murphy, E.K., Peacock, E.E., Reddy, C.H.: The size, mass, and composition of plastic debris in the western North Atlantic Ocean. Marine Pollution Bulletin 2010, 60, 1873-1878.

[5] Thompson, R.C., Olsen, Y., Mitchell, R.P., Davis, A., Rowland, S.J., John, A.W.G., McGonigle, D., Russell, A.E.: Lost at sea: where is all the plastic? Science 2004, 304, 838.

[6] Corcoran, P.L., Biesinger, M.C., Grifi, M.: Plastics and beaches: a degrading relationship. Marine Pollution Bulletin 2009, 58, 80-84.

[7] Rios, L.M., Moore, C., Jones, P.R.: Persistent organic pollutants carried by synthetic polymers in the ocean environment. Marine Pollution Bulletin 2007, 54, 1230–1237.

[8] Moore, C.J.: Synthetic polymers in the marine environment: A rapidly increasing, long-term threat. Environmental Research 2008, 108, 131-139.

[9] Windsor, F.M., Tilley, R.M., Tyler, C.R., Ormerod, S.J.: Microplastic ingestion by riverine macroinvertebrates. Science of The Total Environment 2019, 646, 68-74.

[10] Courtene-Jones, W., Quinn, B., Ewins, C., Gary, S.F, Narayanaswamy, B.E.: Consistent microplastic ingestion by deep-sea invertebrates over the last four decades (1976-2015), a study from the North East Atlantic. Environmental Pollution 2019, 244, 503-512.

[11] IBGE – Instituto Brasileiro de Geografia e Estatística: Cidades. 2018.

[12] Quinn, B., Murphy, F., Ewins, C.: Validation of density separation for the rapid recovery of microplastics from sediment. Analytical Methods 2017, 9, 1491-1498.

[13] Zobkov, M., Esiukova, E.: Microplastics in a Marine Environment: Review of Methods for Sampling, Processing, and Analyzing Microplastics in Water, Bottom Sediments, and Coastal Deposits. Oceanology 2018, 58, 149-157.

[14] Lacerda, A.L.F, Rodrigues, L.S; Sebille, E. V., Rodrigues, F. L., Ribeiro, L., Secchi, E.R., Kessler, F., Proietti, M.C.: Plastics in sea surface waters around the Antarctic Peninsula. Scientific Reports 2019, 9, 3977-3989.

[15] Zhang, H.: Transport of microplastics in coastal seas. Estuarine, Coastal and Shelf Science 2017, 199, 74-86.

[16] Zhou, Q., Zhang, H., Fu, C., Zhou, Y., Dai, Z., Li, Y., Tu, C., Luo, Y.: The distribution and morphology of microplastics in coastal soils adjacent to the Bohai Sea and the Yellow Sea. Geoderma 2018, 322, 201-208.

[17] Andrady, A.L.: Microplastics in the marine environment. Marine Pollution Bulletin 2011, 62, 1596-1605.

[18] Willis, K.A., Eriksen, R., Wilcox, C., Hardesty, B.D.: Microplastic Distribution at Different Sediment Depths in an Urban Estuary. Frontiers in Marine Science 2017, 4, 53-60.

[19] Isobe, A., Iwasaki, S., Uchida, K., Tokai, T.: Abundance of non-conservative microplastics in the upper ocean from 1957 to 2066. Nature Communications 2019, 10, 417.

SAMPLING STRATEGY TO CONDUCT STUDIES ON SPATIAL AND TEMPORAL CONTAMINATION OF MICROPLASTICS IN ESTUARINE ECOSYSTEMS

Mário Barletta

Laboratório de Ecologia e Gerenciamento de Ecossistemas Costeiro e Estuarino (LEGECE), Departamento De Oceanografia (DOCEAN), Universidade Federal de Pernambuco (UFPE), Campus Universitário, 50740-550, Recife – Pernambuco – Brazil; barletta@ufpe.br

Keywords: Fish ontogenetic phase, Feeding Ecology, Microplastics Contamination, Estuary

Abstract

Microplastics (MPs) are present in high densities in estuaries, where they become easily available to the biota. The aim of this study was to describe a methodology developed to study the relationship among the spatio-temporal patterns of habitat utilization, feeding ecology and MPs contamination across the different ontogenetic phases of fishes belonging to different trophic levels and living along the riverine-estuarine-coastal food chain (ecocline) of a tropical environment case study (Goiana Estuary). The water column of Goiana Estuary was examined for the seasonal and spatial variation of MPs (< 5 mm) and their quantification relative to zooplankton and demersal fish species contamination following the same sampling design. The fish species were assigned to different size classes (juvenile, sub-adult and adult). A total mean of $13.7 \times 10^3/m^3$ (Zooplankton: 99%, Ichthyoplankton: 0.6%, microplastics: 0.2%) were captured in the seston. The highest amounts of MPs were observed during the late rainy season. The density of MPs in the water column determines their bioavailability to plankton feeders, and then to larger predators. This process possibly promotes the transfer of MPs between trophic levels. Polyamide (Nylon) and polyethylene blue fibres were the most frequent types of MPs found in the estuary. Their presence could possibly be linked to nets and cables used in fisheries and other maritime operations. Juveniles of socio-economically important species (*Cynoscion acoupa* – Acoupa weakfish and *Centropomus undecimalis* - Snooks) occur in the upper estuary, an important nursery ground during the early rainy season. Sub-adults also inhabit the upper estuary, a rich feeding ground, where marine predators are absent and competition is reduced. When river runoff increases, Acoupa weakfish and Snooks move to adjacent coastal areas. During this time of the year, species of these genera spawn close to the entrance of the estuarine ecosystem. During the end of the dry season, the juveniles of *C. acoupa* and *C. undecimalis* use the upper and middle estuary as nursery. Studies on MPs distribution in relation to spatial and temporal variation of the fauna and environmental factors, which influence the movements of the marine biota, are increasing in quantity and quality. If these sampling strategies are replicated in other estuaries, comparisons could be made. Standard protocols for sampling, extraction, enumeration and classification of MPs ingested by fishes have been developed and are presented here in order to enable worldwide comparisons. Standardized sample designs and laboratory procedures are an important strategy in order to establish comparisons among different sites. Also, it facilitates comparisons along time periods when studying the same environment, which is important when assessing the effectiveness of managerial measures taken to abate pollution.

1 Introduction

Estuaries are very productive ecosystems that offer a number of environmental services and resources. However, anthropogenic interference usually results in loses that affect directly the fauna across different trophic levels. One of the most conspicuous forms of interference is pollution by plastic waste. Once feral, plastics fragment into smaller particles that eventually reach the size of MPs (< 5mm). In addition, studies revealed that fisheries (including the wear and tear of gear) and domestic effluents (fibres) produce and dispose of a large amount of microfilaments that also reach estuaries. The main sources of MPs to estuaries are river basins, followed by autochthonous sources [1]. From there, they eventually reach the sea.

MPs present high densities in estuaries, where they become easily available for the biota. Among these organisms are fish, which have different feeding habits and habitats uses, depending on species, ontogenetic phase and seasonal influences. If one of the phases within the life cycle of a fish species lives or uses an estuarine habitat when it is highly contaminated by MPs, the chances of interactions increase significantly. Interactions include the direct and indirect ingestion of MPs and, therefore, must consider the shifting patterns of feeding habits along ontogeny. When feeding, fish may pray on dietary items previously contaminated [2].

Contamination usually increases towards the top of the food chain, and higher trophic levels are more contaminated by MPs, especially developed phases that prey on larger items [3]. The trophic transfer of MPs along food chains of the Goiana Estuary is already being studied, and top predators show higher contamination levels in their later phases that prey on whole fishes. The ingestion of MPs by species of economic and social interest might become a public health issue, if MP shave adsorbed pollutants available in the environment onto their surfaces before ingestion.

The aim of this study was to describe a methodology developed to study the relationship among the spatio-temporal patterns of habitat utilization, feeding ecology and MPs contamination across the different ontogenetic phases of fishes in different trophic levels of a riverine-estuarine-coastal food chain and the ecocline of a model tropical environment. For this purpose, Goiana Estuary was used as a case study.

2 Materials and Methods

2.1 Study area

The Goiana River basin has a total drainage area of 2,900 km^2 [4]. The network of small channels also serves as receptor of domestic and industrial effluents and wastes along its short and low-volume flow. The population served by this basin for water is of approximately 500,000 inhabitants. Average water flow is around 11 m^3/s^1 (0.5 to 25 m^3/s). The last 17 km of its course is under the influence of tidal cycles, characterizing the estuary (Figure 1).

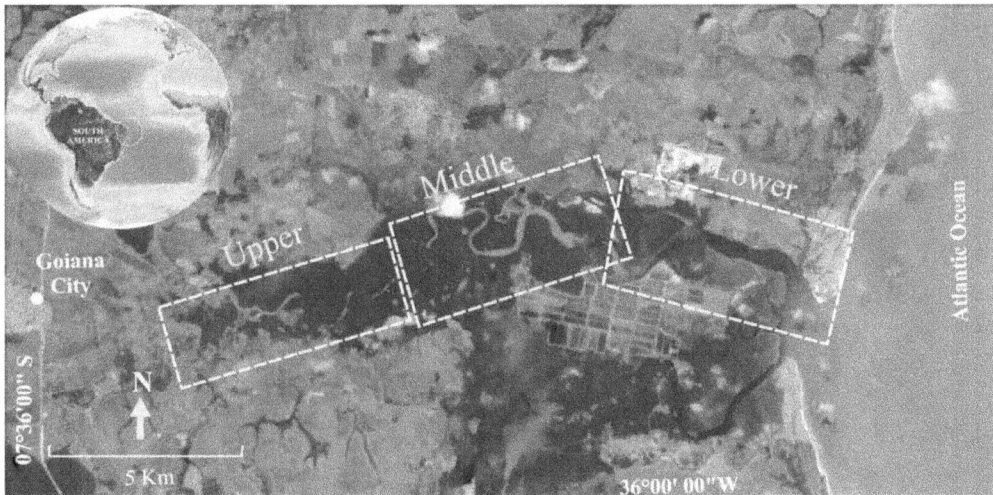

Figure 1: Estuary of Goiana River. The main channel was divided in three areas (Upper, Middle and Lower) according to the gradient of salinity and geomorphology of the estuary.

2.2 Sampling Strategies

The water column of the model estuary (Figure 1) was examined for the seasonal and spatial variation of MPs (< 5 mm) and their quantification relative to seston (*e.g.*, MPs and zooplankton) and demersal fish species contamination following the sampling design [5]. In the absence of obvious sources, these MPs are inferred to be mostly the result of the degradation of larger items, accumulated in the flooded forest and other estuarine habitats.

2.2.1 Seston

The samples conducted to study the seston were 3 superficial (0 – 1 m) and 3 bottom (3 - 6 m) water sample replicates taken monthly (12 months) in each portion of estuary (Figure 1) by towing a conical plankton net (300 μm; Ø 0.6 m; 2 m long) for 15 min at average speed of 2.7 knots [1]. The volume filtered per tow was calculated using a flowmeter. Fish larvae, fish eggs and MPs were totally separated from the bulk sample and their counts per unit were converted to a standard volume of 100 m^3.

2.2.2 Demersal fishes

Six replicate trawls were made per month in each estuarine habitat of the main channel (upper, middle and lower) with an otter trawl net. The net was 8.72 m long with a mesh size of 35 mm in the body and 22 mm in the cod-end. The length of ground rope was 8.5 m, and the head rope was 7.1 m long. To obtain a representative sample of all the fish sizes, a cover with a smaller mesh size (5 mm) was used over the cod-end [6].

The fish species were assigned to different size classes (juvenile, sub-adult and adult) based on 2 criteria. The first criterion was length at first maturation (L_{50}) calculated [7] and used to distinguish sub-adults from adults (Figure 2A). The second criterion was the inflection point of the length–weight curve, which was used to distinguish the juveniles from sub-adults (Figure 2B). The stomach contents of each size class for each area (upper, middle and lower estuaries) and season (early and late dry seasons, late and rainy seasons) was examined using a stereomicroscope (Zeiss x 50). All dietary and non-dietary items (*e.g.*, plastics, nylon) were separated and identified.

3 Results and Discussion

A total mean density of ~$13.7 \times 10^3/m^3$ items was captured in the seston during the sampling period (Table 1) [1, 8]. From this total, ~99% was composed of Zooplankton. Ichthyoplankton and MPs contributed with 0.6 and 0.2, respectively.

The highest amount of MPs was observed during the late rainy season, when the environment is under the influence of the highest river flow, which induces the runoff of plastic fragments to the lower estuary (Figure 3). In the upper estuary (surface) and in the lower estuary (bottom) during the late rainy season, the concentration of MPs was higher than the concentration of fish larvae.

Microfilaments (0.4/100 m^3) correspond to only 0.01% of total seston observed in the main channel of Goiana Estuary (Table 1). However, despite the small concentration of microfilaments in seston composition, the microfilaments were the most common MPs found in fishes gut contents, independent from ontogenetic phase and/or trophic level [5]. The concentration of the microfilaments in the water column showed the highest values in the surface of the lower estuary, principally during late rainy season and late dry season (Figure 4). The upper estuary showed more concentration of microfilaments on the bottom during the dry season. During the early dry season and early rainy season, juveniles of Snooks (*Centropomus* spp) and Acoupa weakfish (*Cynoscion* sp) use this portion of the estuary as a nursery ground [2]. The juveniles of these species prey principally on zoobenthos invertebrates in the nursery ground. It suggests that the high concentration of microfilaments during this time of the year increases the chance of contamination of the juveniles of these species.

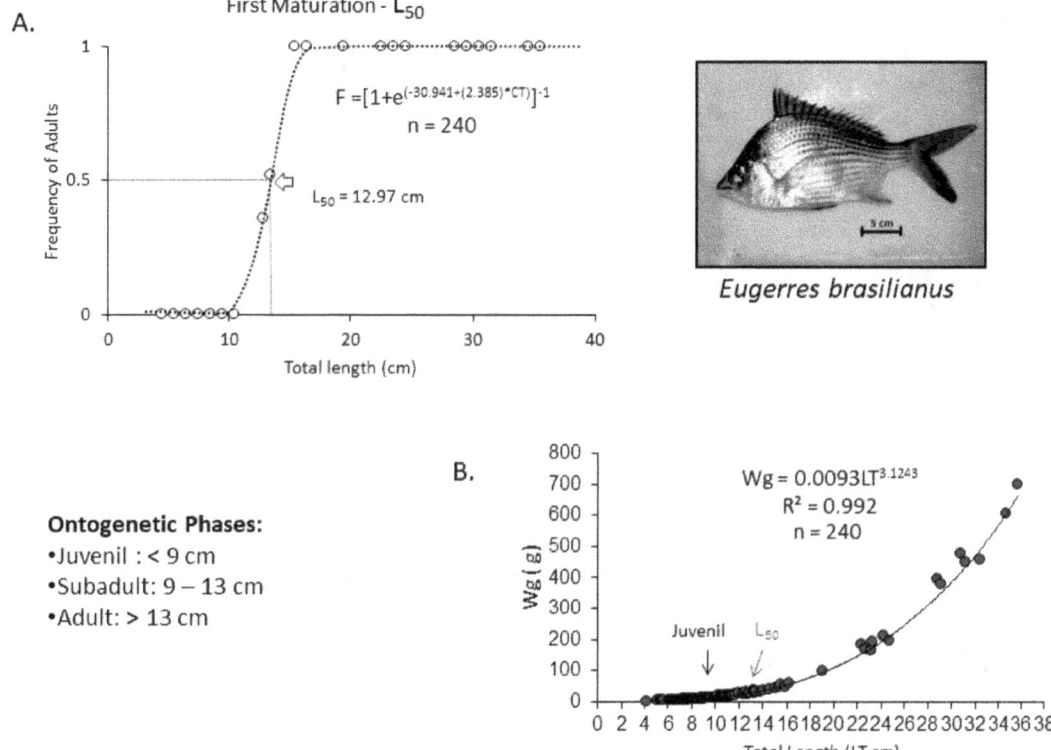

Figure 2: (A) Relative frequency of adults (lines and arrow in maturation curve indicate length at first maturity [L_{50}]), and (B) Length–weight relationship (arrows indicate size at inflection point) for *Eugerres brasilianus* (modified from [9]).

However, in stratified estuaries, the salt wedge and its seasonal variations can be a factor in MPs density differentiation (Figure 5). All of the above studies were conducted according to a strict sample design that allowed the results to be statistically treated to identify spatial and temporal variations in ecological variables, including the ingestion of MPs by fish [5]. The adherence to sampling designs that can positively identify spatio-temporal patterns in the distribution of ecological variables and pollutants, such as MPs, is important for the suggestion of the most probable sources and the ingestion or entanglement risks [10]. Polyamide (Nylon) and polyethylene blue fibres were the most frequent types of microfilament found in the Goiana Estuary (Figure 4). Their presence could possibly be linked to nets and cables used in fisheries and other maritime operations.

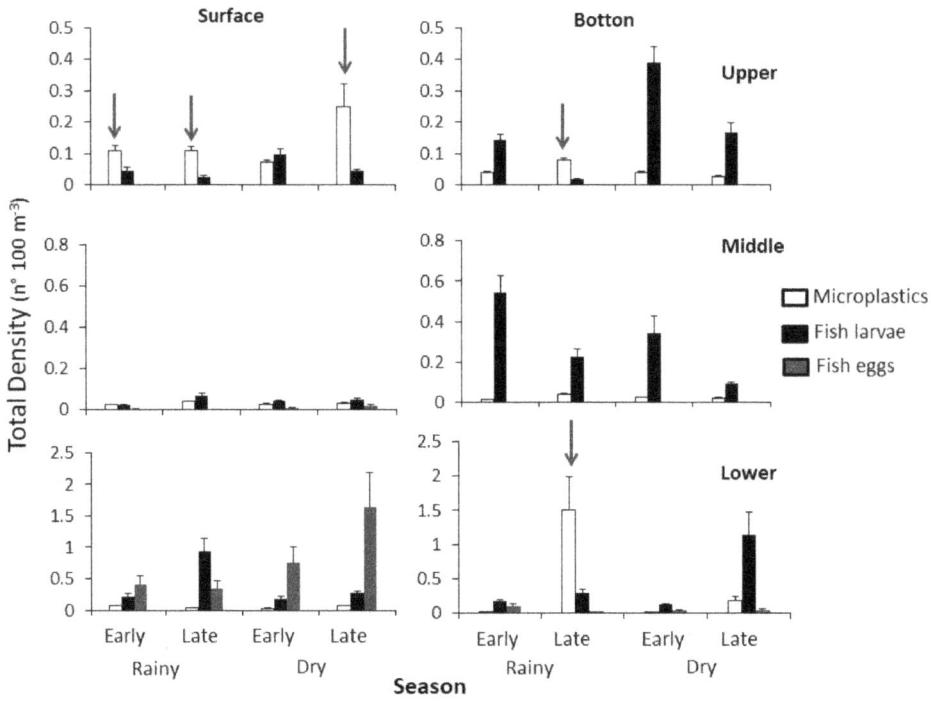

Figure 3: Seasonal fluctuations of total mean density (mean ± SD) of MPs, fish larvae and fish eggs in each area of the Goiana Estuary (Upper, Middle and Lower)

The near-absence of plastic pellets in these works suggests that MPs in this environment are secondary ones, resulting from the fragmentation of larger items, either along the river basin or within the estuary. Marine sources are less likely or have a weaker signal. It was not possible to determine and to quantify the real damage caused by MPs to individuals [8], but researchers can identify possible threats at the population level [2, 3, 9-11].

The percentage of the fish species, which was contaminated with blue microfilaments is Catfishes 20% (*Cathorops spixii, C. agassizii* and *Sciades herzbergii*) [10], Mojaras 13% (*Eusinostomos melanopterus, Eugerres brasilianus, Diapterus rhombeus*) [9] and Drums 8% (*Stellifer brasilianus* and *S. stellifer*) [11].

In the Goiana Estuary, juveniles of many socio-economically important species (*e.g., Cynoscion acoupa* and *Centropomus undecimalis*) of the Western Tropical Atlantic occur mainly in the upper estuary. This portion of estuary is an important nursery ground during the early rainy season for juveniles of *C. acoupa* (Figure 5), and during early dry season for juveniles of *C. undecimalis* (Figure 6). Sub-adults of both species also inhabit the upper estuary as rich feeding ground, where marine predators are absent and competition reduced [2, 3, 12].

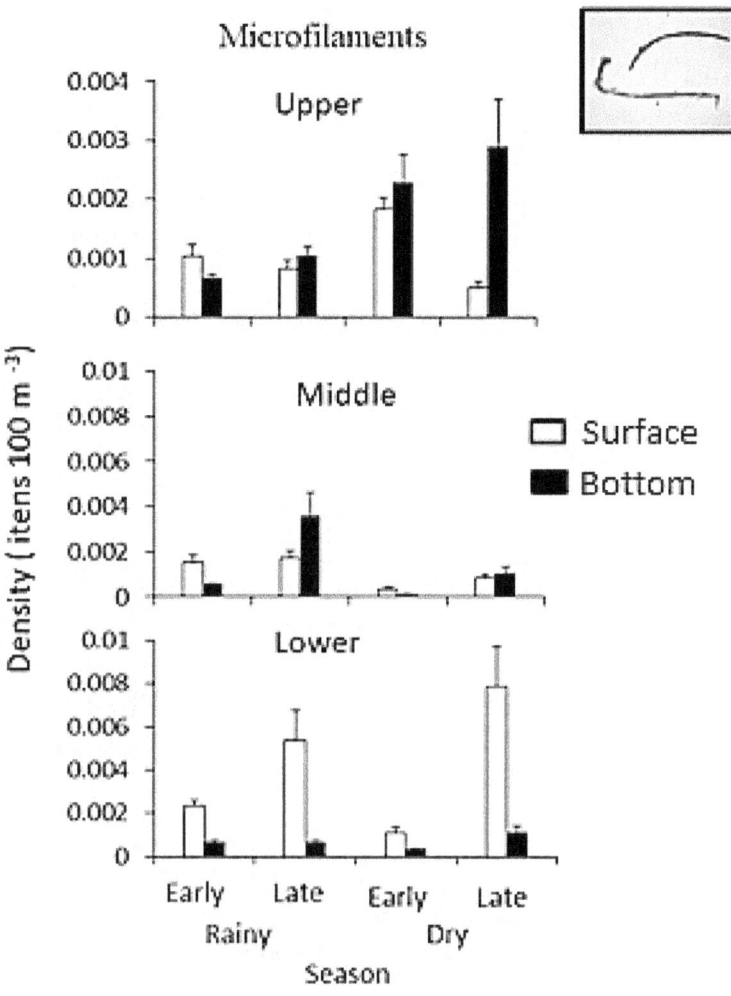

Figure 4: Fluctuation of microfilaments density (surface: white; bottom: black) (mean ± SD) in function to the variables season (early and late rainy; early and late dry) and area (upper, middle and lower)

Adults of *C. acoupa* (Acoupa weakfish) are one of the main predators inhabiting adjacent coastal areas [12]. The juvenile and sub-adults utilise principally the main channel of the estuarine ecosystem to complete their life cycle. For that reason, this species is classified as estuarine dependent [13].

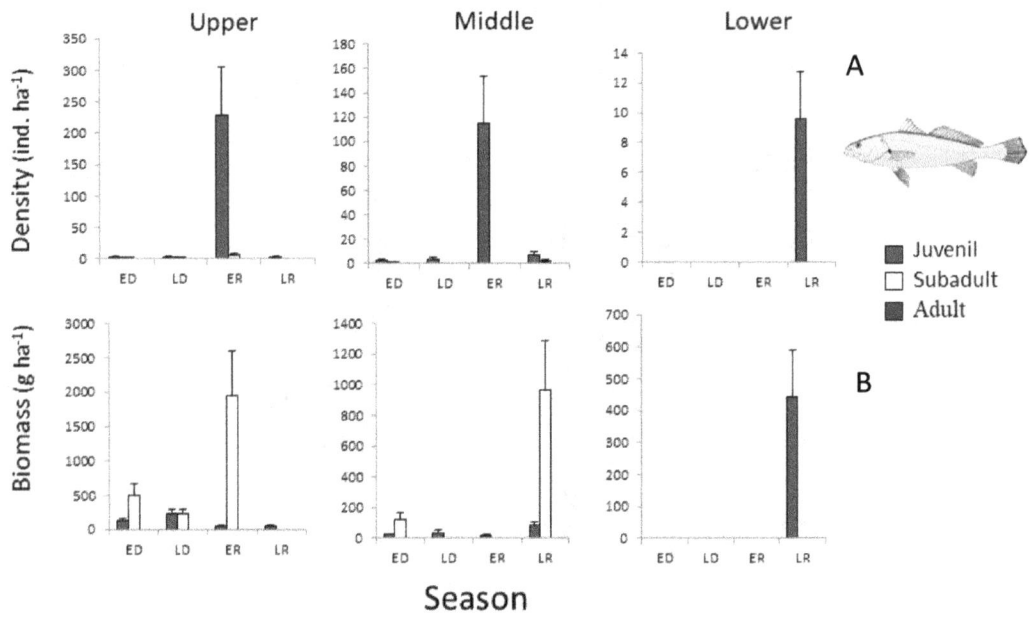

Figure 5: Seasonal fluctuation of density (mean ± SD) (A) and biomass (B) of *Cynoscion acoupa* (juvenile: blue; sub-adult: white; adult: red) in different portion of estuary (upper, middle and lower); Season is represented by ED: early and LD: late Dry, and ER: early and LR: late rainy season.

Species of Snook (*e.g.*, *C. undecimalis*) is semi-anadromous [14]. During the dry season, adults move into the upper portion of the estuary (Figure 6). When river runoff increases because of the rainy season, Snooks move into the coastal areas adjacent to the estuary. During this time of the year, species of these genera spawn close to the entrance of the estuarine ecosystem. Larvae drift into the estuarine mangrove forest, where they remain until reaching the juvenile stage [8]. During the beginning of the dry season, the juveniles of *C. undecimalis* use the upper and middle estuary as nursery (Figure 6). During the end of rainy season, specimens of *C. acoupa* and *C. undecimalis* concentrate in the coastal areas adjacent to the estuary, where the artisanal fishery captures them. Therefore, studying the effects of non-natural food items on *C. acoupa* and *C. undecimalis* is fundamental to understand MPs role in the ecology of the estuary [5]. For all ontogenetic phases of both species, individuals contaminated principally by blue microfilament were detected (Figure 7). Juveniles of both species showed microfilaments in the gut content in the nursery area (upper estuary during the early dry season). However, the adults showed the highest concentrations principally in the lower estuary.

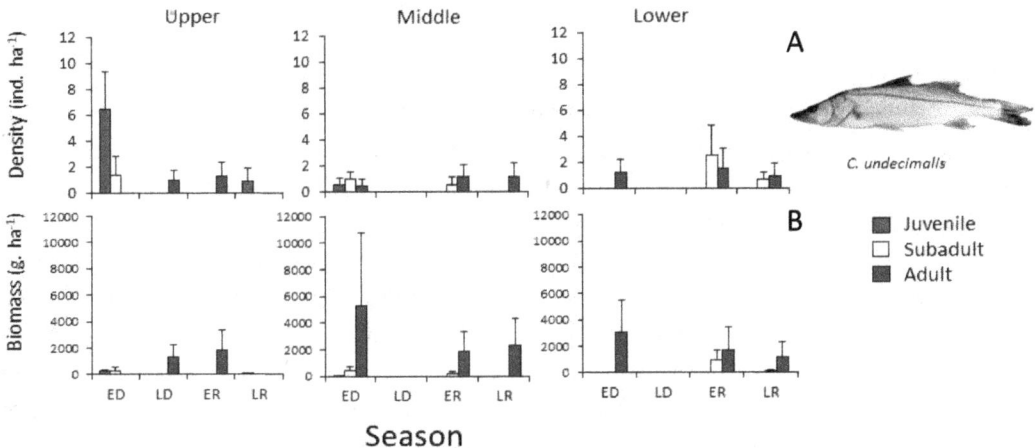

Figure 6: Seasonal fluctuation of density (mean ± SD) (A) and biomass (B) of *C. undecimalis* (juvenile: blue; sub-adult: white; adult: read) in relation to the variable reach of estuary (upper, middle and lower). Season is represented by ED: early and LD: late Dry, and ER: early and LR: late rainy season.

Studies on MPs distribution in relation to spatial and temporal variation of the fauna and the environmental factors, which influence the movements of the marine biota, are increasing in quantity and quality [1-3, 8, 11, 12, 15]. If these sampling strategies could be replicated in other estuaries, comparisons would be made possible [5]. Then, managers of different areas/regions of the world could make plans to overcome the problem of environmental contamination with MPs.

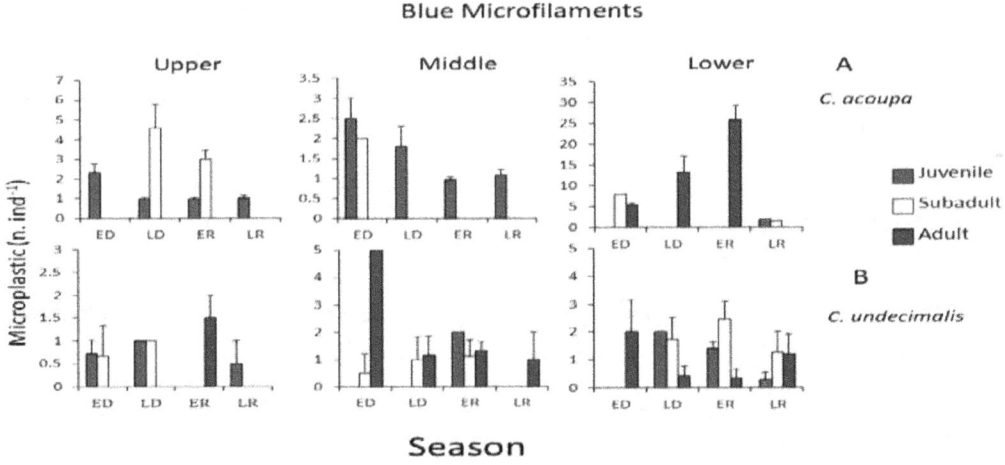

Figure 7: MPs contamination (mean ± SD) for different ontogenetic phase (juvenile: blue, subadult: white; adult: read) of *Cynoscion acoupa* (A) and *Centropomus undecimalis* (B) in different portions

of estuary (upper, middle and lower). Season is represented by ED: early and LD: late Dry, and ER: early and LR: late rainy season.

4 Conclusions

Standard protocols for sampling, extraction and enumeration of MPs ingested by fishes enable worldwide comparisons as well as temporal assessments of the same site in order to detect changes. Although several studies have focused on the contamination of fishes by plastic debris, few attempts have been made to understand spatio-temporal patterns of availability and ingestion of MPs. Both the distribution patterns of fishes and MPs availability vary with the spatial and seasonal variability of environmental factors within tropical and sub-tropical estuaries. Any investigation aiming at determining MPs (and/or other pollutants) impact in estuarine environments must include the role of the estuarine ecocline on fish ecological behaviour and on their encounter rate with MPs. This approach is important to detect, which environmental variables are associated with patterns of MPs exposure and ingestion through the life cycle of fish species, in addition to changes in fish and all fauna patterns of use within the estuary.

5 Acknowledgements

The author acknowledges the support of DAAD and Exceed Swindon project, which made the participation at the event possible. In addition, the author thanks the financial support from Conselho Nacional de Desenvolvimento Científico e Tecnológico through grants CNPq-Proc.405818/2012-/COAGRE/PESCA, and Fundação de Apoio à Pesquisa do Estado de Pernambuco (FACEPE) through Grant FACEPE/APQ-0911-108/12. M.B. is a CNPq Fellow.

6 References

[1] A.R.A. Lima, M.F. Costa, M. Barletta: Distribution patterns of microplastics within the plankton of a tropical estuary. Environmental Research 2014, 132, 146-155.

[2] G.V.B. Ferreira, M. Barletta, A.R.A. Lima, S.A. Morley, M.F. Costa: High intake rates of microplastics in a Western Atlantic predatory fish, and insights of a direct fishery effect. Environmental Pollution 2018, 236, 706-707.

[3] G.V.B. Ferreira, M. Barletta, A.R.A. Lima, S.A. Morley, M.F. Costa: High intake rates of microplastics in a Western Atlantic predatory fish, and insights of a direct fishery effect. Science of The Total Environment 2019, 236, 706-717.

[4] M. Barletta, M.F. Costa: Living and Non-living Resources Exploitation in a Tropical Semi-arid Estuary. ICS Proceedings, Journal of Coastal Research 2009, SI 57, 371-375.

[5] M. Barletta, A.R.A. Lima, M.F. Costa: Distribution, sources and consequences of nutrients, persistent organic pollutants, metals and microplastics in South American estuaries. Science of The Total Environment 2019, 651, 1199-1218.

[6] M. Barletta, A. Barletta-Bergan, U. Saint-Paul, G. Hubolt: The role of salinity in structuring the fish assemblages in a tropical estuary. Journal of Fish Biology 2005, 66, 45–72.

[7] D.D.S Lewis, N.F. Fontoura,: Maturity and growth of *Paralonchurus brasiliensis* females in southern Brazil (Teleostei, Perciformes, Sciaenidae). Journal of Applied Ichthyology 2005, 21, 94–100.

[8] A.R.A. Lima, M. Barletta, M.F. Costa: Seasonal distribution and interactions between plankton and microplastics in a tropical estuary. Estuarine, Costal and Shelf Science 2015, 165, 213-225.

[9] J.A. Ramos, M. Barletta, M.F. Costa: Ingestion of nylon threads by Gerreidae while using a tropical estuary as foraging grounds. Aquatic Biology 2012, 17, 29-34.

[10] F.E. Possatto, M. Barletta, M.F. Costa, J. Ivar do Sul, D.V. Dantas: Plastic debris ingestion by marine catfish: An unexpected fisheries impact. Marine Pollution Bulletin 2011, 62, 1098-1102.

[11] D.V. Dantas, M. Barletta, M.F. Costa: The seasonal and spatial patterns of ingestion of polyfilament nylon fragments by estuarine drums (Sciaenidae). Environmental Science and Pollution Research 2012, 19, 600-606.

[12] G.V.B. Ferreira, M. Barletta, A.R.A. Lima, D.V. Dantas, A.K.S. Justino, M.F. Costa: Plastic debris contamination in the life cycle of Acoupa weakfish (*Cynoscion acoupa*) in a tropical estuary. ICES Journal of Marine Science 2016, 73, 2695-2707.

[13] M. Barletta, S.J.M. Blaber: Comparison of fish assemblages and guilds in tropical habitats of the Embley (Indo-West Pacific) and Caeté (Western Atlantic) estuaries. Bulletin of Marine Science 2007, 80(3), 647–680.

[14] D.V. Dantas, M. Barletta: Habitat use by *Centropomus undecimalis* in a rocky area of estuarine beach in North-East Brazil. Journal of Fish Biology 2016, 89, 793–803.

[15] J. D. Silva, M. Barletta, A.R.A. Lima, G.V.B. Ferreira: Use of resources and microplastic contamination throughout the life cycle of grunts (Haemulidae) in a tropical estuary. Environmental Pollution 2018, 242, 1010-1021.

DETECTION AND IDENTIFICATION OF MICROPLASTIC MATERIALS WASTE IN WATER SYSTEMS

S. El Hajjaji[1], J. Mabrouki[1], C. Bakkouche[1], A. Dahchour[2]

[1]*Laboratory of Spectroscopy, Molecular Modeling, Materials, Nanomaterials, Water and Environment, CERNE2D, Mohammed V University in Rabat, Faculty of Science, AV Ibn Battouta, BP1014, Agdal, Rabat, Morocco; souad.elhajjaji@um5.ac.ma, hajjajisouad@yahoo.fr*

[2]*Agronomic and Veterinary Institute Hassan II, Rabat, Morocco*

Keywords: Microplastics, freshwater systems, detection, identification, waste

Abstract

Microplastics (MPs) are particles smaller than 5 mm, object of an increasing number of investigations that demonstrates being this issue concern of all compartments and lattices of the earth and featuring developing doubts about their poisonous quality. In this specific circumstance, examination of the procedures utilized for the identification (ID) and measurement of MPs in water is very relevant. Plastic particles were outwardly recognized, extricated, checked and gauged, at that point the concoction structure of one piece of them distinguished by infrared spectroscopy (FT-IR ATR). Plastic particles were found in all samples, the mass being from the discontinuity of bigger plastic articles and particularly plastic sacks or bundling. The greater part of the dissected particles comprised of PET, PE and PVC. This exploratory research requires a developing of such diagnostic strategies on tests, on a bigger scale or on smaller size portions.

1 Introduction

Population and economic growth around the globe has consequences for the environment and water bodies. Thus, the water resources are subjected to a strong pressure exerted by the anthropic activity (agriculture, industry, breeding, fishing, domestic, hospitals, etc.). Anthropogenic activities of socio-economic nature coupled with those of natural processes (soil erosion, precipitation, evaporation, runoff of river water) accelerate the degradation of surface water resources [1].

MPs pollution is considered as a worldwide problem and direct threat to the environment and human health through the potential to transport the contaminants and their impacts on biodiversity [2–7]. According to natural dynamic force of the ocean and sea surface, and the large area involved, it is difficult to predict, where it can accumulate, and thus will cause the greatest impacts. Importantly, the incidence and impact of plastic debris on surface water remain largely unknown. Nowadays, the contamination of the hydrosphere by plastic waste has become a worldwide emerging concern, posing serious threats to humans and their natural environment due to their ubiquitous occurrence, bioavailability and ability to carry toxic chemicals [8-10]

This article presents the synthesis of fluxes of anthropogenic pollutants and their impacts on aquatic resources and human health. This literature review is, therefore, a step in the constitution of a database that can better guide research on the influence of MPs on water resources.

2 Freshwater, different types of MPs and pollutants

2.1 Definition, history and type of MPs

Microplastics (MPs) have been defined as polymer particles [9] or tiny plastic pieces [11], whose sizes do not exceed 5 mm [12]. This definition is adopted by several international bodies such as the Food and Agriculture Organization of the United Nations (FAO) [13], the Convention for the Protection of the Marine Environment of the North-East Atlantic (OSPAR) [14], and the group of Experts on the Scientific Aspects of Marine Environmental Protection (GESAMP) [5]. Other definitions consider MPs as particles having a size less than 1 mm [15, 16], or less than 2 mm [17], and also between 2 and 6 mm, or even less than 10 mm [18].

MPs are found in a variety of products ranging from cosmetics over synthetic garments to plastic bags and bottles that easily enter the environment as a result of pollution in form of waste, and remain both in sea water and freshwater [19]. MPs are divided into two types:

- *Primary MPs*: microbeads incorporated into cosmetics or personal care [20] and personal hygiene products such as peels, shower gels or toothpastes to increase the cleaning power by their abrasive properties [21]. Plastic pellets (or noodles) used in industrial manufacturing and plastic fibers used in synthetic textiles (e.g., nylon) enter the environment directly through various channels, such as household sanitation systems, unintentional losses caused by spills during manufacture or transportation, or abrasion during washing (e.g., washing of garments made of synthetic textiles).

- *Secondary MPs*: are formed from the breakdown of larger plastics. This usually happens, when larger plastics undergo alteration, e.g., exposure to sun light and natural mechanical forces such as waves [21].

2.2 Effects of MPs

Recently, MPs have become global contaminants concern for human and ecological health. They are not biodegradable and are found in a variety of environments, such as oceans and freshwater ecosystems, where they accumulate and persist, and remain bioavailable because of their small size for thousands of species from almost every trophic level because they are often mistaken for food [22].

Globally, several studies have been conducted on the potential toxic effects of MPs. They revealed that MPs can have a variety of adverse physical effects on humans and living organisms[19] that ingest plastic debris [23]. As a result, aquatic species ingest MPs floating on the water surface, confusing them with food, which means that they consume less food and, therefore, have less food and energy to perform their vital functions, which can lead to neurological and reproductive toxicity. MPs are suspected of climbing marine food chains from zooplankton and small fish to large marine predators.

MPs are also a source of air pollution for their presence in dust and fibrous particles suspended in the air. The effects of inhalation of MPs on health are unknown. MPs are of increasing concern, because they pose a threat to aquatic species and humans. They do not only accumulate in the environment but, due to their absorption potential, they can also contribute to the spread of micropollutants in the environment [20].

Until recently, there was very little information on transfer of MPs to humans. Since they are the ultimate consumer of the food channel, the introduction of MPs into humans seems possible through the consumption of aquatic products containing plastics [24].

3 Methods to detect and to identify MPs

3.1 Detection of MPs

Due to lack of harmonization, many methodologies have been developed to detect and to quantify MPs in water, sediment and at a lesser extent in organisms. The first tests for MPs in fish made at INERIS (spell) were made by direct observation of the stomach contents using a binocular loupe. It soon became apparent that a specific methodology was needed to facilitate the detection and identification of MPs. Thus, a two-step protocol has been developed. This combines a densitometric separation carried out with a saturated solution of sodium sulphate (Na_2SO_4) to a coloration of the supernatant particles with "Nile Red". This methodology makes it possible to keep MPs intact and facilitates their observation by isolating them from the organic matter present in the digestive tubes of fish. It allows, unlike direct observation, the detection of microbeads.

3.2 Identifying, measuring and monitoring MPs

Unlike other pollutants, the identification of MPs seems to be a challenge, either they would be in wastewater or oceans, because of the lack of clear definition, the diversity of their sizes and their composition, making it difficult to develop an identification method using only one analytical technique [10], as well as the comparison of the results between different studies. Therefore, the combination of more than two techniques is widely used. Generally, the identification of MPs is carried out in two stages, namely a physical characterization of the potential plastics (for example, microscopy), followed by a chemical characterization of the isolated particles in mixtures of inorganic and organic residual particles after the steps of extraction and cleaning (e.g., spectroscopy) for the confirmation of plastics [10].

However, to identify MPs pieces and to determine their chemical composition, the most credible method is Fourier Transform Infrared Spectroscopy (FTIR) [21]. However, this method requires much more time than a simple microscopic examination and its detection limit is around 20 microns only. Nevertheless, microscopic identification alone has a high risk of producing false results when analyzing small MPs [10]. For this, another method of spectroscopic confirmation that completes the FTIR is Raman spectroscopy, which makes it possible to detect particles of 1-2 µm in diameter, to be certain that the MPs, including the microfibers, are MPs (for example, no cotton or cellulose fibers).

3.3 Sampling and analysis of MPs

The methods usually used for the analysis of MPs in aquatic sediments were adapted to the nature of the sediments and the laboratory experience, as well as to the requirements. While MPs studies usually focus on relatively low-density plastics (e.g., PP, PE and expanded PS), it has been chosen here to carry out a broader search for all plastics likely to be found in sediments, which necessitated the development of a specific methodology. The samples supplied with a volume between 2 and 5 L, were first homogenized and 1 L was taken [25-26]. Each sample was rinsed with large volumes of water above sieves, stacked to separate into 4 classes:

- 5 mm, boundary between mesoplastics and MPs ;
- 5 to 1 mm, boundary between small and large MPs, to distinguish between two commonly used definitions of MPs [27].
- 1 mm to 300 µm, lower limit used in practice in most comparable studies;
- < 300 µm, preserved but not analyzed here.

Only fractions greater than 300 µm were studied, firstly because this limit is the one generally used for studies in aquatic environments [28], and secondly for practical reasons, handling and even more identification of plastics to smaller sizes are proving uncertain.

4 Monitoring MPs in freshwater systems

4.1 MPs transfer to the food chain

In Europe, 30% of plastic objects are recycled and transformed into other products, 40% are burned to produce energy and, therefore, 30% still go to landfills [29]. Marine animals undergo the presence of plastic, which is on the edge of the coast (50-80% of waste). Scientists have discovered more than 260 different animal species that have ingested plastic: turtles, birds, mammals and fish [6, 30], while ingestion of marine debris is known to particularly affect turtles and seabirds

The greatest threat of ingestion occurs, when it blocks the digestive tract or fills the stomach, leading to a potential death of these species [31]. Since humans are at the end of the food chain, they can ingest MPs by consuming seafood products. The exposed of MPs might contain chemical compounds and plastic additives like phthalates [32-34]. Indeed, a number of compounds associated with plastics, including phthalate and bisphenol A may cause health disorders in humans. But plastics also contain other chemical pollutants that could be harmful but are still subject to study [35, 36].

Given their persistence in the environment, one of the solutions to tackle the MPs pollution would be to develop and to produce plastics degradable under environmental conditions [37–38].

4.2 MPs in Morocco

Morocco ranks at the 18th place on the list of top countries that mismanage plastic waste, and it is estimated to generate 0.05-0.18 million t/yr of plastic debris, a rank equivalent to what the 23 coastal countries of EU would score together. Plastic contamination is a growing environmental problem in marine systems. Pieces of plastic in small scales, MPs (particles < 5 mm), have reached high abundance (e.g., 100,000 particles/m^3) of water and sediment, and interact with organisms and the environment in a variety of ways. Morocco coasts are 3,500 km long with a Mediterranean coast line of 500 km and Atlantic coast line of 3.000 km.

In a study on the pollution of Atlantic Ocean, authors reported an average MPs abundance estimated at 1.15 ± 1.45 particles/m^3, which is lower than it was reported for sub-surface waters across the world. The most abundant polymer types were polyester (49%) and blends of polyamide or acrylic/polyester (43%). Fibres (94 %) were also the predominant type of MPs.

Morocco is aware about his fish export and production such as canned sardine. In a recent study about this species, 20 marketed brands from 13 countries over 4 continents are investigated for the potential presence of MPs and mesoplastics. Plastic particles were absent in 16 brands, while between 1 and 3 plastic particles per brand were found in the other 4 brands.

The negative impacts of MPs were investigated on two marine species *Auxis thazard* (AT) and *Diplodus cervinus* (DC), spread on the northern coast of the Mediterranean Sea in Morocco. Various marine litters have been found by fishermen using sea-nets that affect fishing yields and thus the financial income of fishermen. Out of 30 samples of each species studied, only 3 and 4 samples showed MPs ingestion for AT and DC, respectively. Thousands of small pieces from marine litter were reported as a result of degradation process. This indicates that 10% of fish ingests plastic directly or indirectly.

Recent studies of freshwater systems suggest the presence of MPs, and interactions could have the same range as those observed in marine systems. Studies on the accumulation and effects of plastics in freshwater and terrestrial systems are much less than in marine systems. Differences between freshwater and marine systems include closer proximity to point sources in freshwaters, smaller freshwater system sizes, and spatial and temporal differences in particle mixing/transport. These differences between marine and freshwater systems can lead to following differences:

- Development of methods to detect, to identify, to measure, and to monitor MPs in freshwater systems.
- Influence of wastewater treatment plant discharges on MPs concentrations in surface water.
- Their transportation routes and factors affecting distributions; the type of MPs present. The extent and relevance of their impacts on aquatic life.
- The most effective degradation process on plastics on river or exposed to air and sunlight.

It is also not known, how MPs could move from freshwater to terrestrial ecosystems, and if and how they can affect human health. A related study will be conducted on Bouregrag River in Rabat (Morocco).

5 Conclusions

Through this bibliographic synthesis, the data collected made it possible to carry out an analysis of the MPs in freshwater. Water pollution has repercussions on the lives of aquatic ecosystems, the food web and the health of the population. MPs are becoming a main source of pollution with negative impacts on agriculture and industrial activities. The methods of analysis and identification of MPs in water are poorly studied. All these results show the acuity of the problem caused by human activity on freshwater. It is important that all MPs are characterized and their impacts investigated in order to limit pollution.

5 Acknowledgements

The authors would like to thank EXCEED Swindon project and DAAD (German Academic Exchange Service) for their support to participate at the Regional Workshop on "Microplastics in the Water Environment", from 19-21 August 2019 at Koh Samui, Thailand.

6 References

[1] S. Aw, B. N'goran, S. Siaka, B. Parinet: Intérêt de l'analyse multidimensionnelle pour l'évaluation de la qualité physico-chimique de l'eau d'un système lacustre tropical: cas des lacs de Yamoussoukro (Côte d'Ivoire). J. Appl. Biosc., 2011, 38, 2573 – 2585.

[2] A. Alshawafi, M. Analla, E. Alwashali, M. Ahechti, M. Aksissou: Impacts of Marine Waste, Ingestion of Microplastic in the Fish, Impact on Fishing Yield, M'diq, Morocco. Int. J. Mar. Biol. Res., 2018, 3(2), 1–14.

[3] J.G.B. Derraik: The pollution of the marine environment by plastic debris: a review. Mar. Pollut. Bull., 2002, 44(9), 842–852.

[4] Y. Mato, T. Isobe, H. Takada, H. Kanehiro, C. Ohtake, T. Kaminuma: Plastic resin pellets as a transport medium for toxic chemicals in the marine environment. Environ. Sci. Technol., 2001, 35(2), 318–324.

[5] GESAMP: Sources, fate and effects of microplastics in the marine environment: Part two of global assessment. International Maritime Organization, London, 2018, 5(3), 375–386; ISSN: 1020-4873.

[6] A.L. Andrady: Microplastics in the marine environment. Mar. Pollut. Bull., 2011, 62(8), 1596–1605.

[7] P.L. Corcoran, Benthic plastic debris in marine and fresh water environments. Environmental Science: Processes & Impacts, 2015, 17(8), 1363–1369. https://doi.org/10.1039/C5EM00188A

[8] L. Frère: Les microplastiques : une menace en rade de Brest. École Doctorale des Sciences de la Mer. Université de Bretagne Occidentale, 2017.

[9] S.M. Mintenig, I. Int-Veen, M.G.J. Löder, S. Primpke, G. Gerdts: Identification of microplastic in effluents of waste water treatment plants using focal plane array-based micro-Fourier-transform infrared imaging, Water Res., 2017, 108, 365–372.

[10] W.J. Shim, S.H. Hong, S.E. Eo: Identification methods in microplastic analysis: a review. Anal. Methods, 2017, 9(9), 1384–1391.

[11] A. Dyachenko, J. Mitchell, N. Arsem: Extraction and identification of microplastic particles from secondary wastewater treatment plant (WWTP) effluent. Anal. Methods, 2017, 9(9), 1412–1418.

[12] L. van Cauwenberghe, A. Vanreusel, J. Mees, C.R. Janssen: Microplastic pollution in deep-sea sediments. Environ. Pollut., 2013, 182, 495–499.

[13] A. Lusher, P. Hollman, J. Mendoza-Hill: Microplastics in fisheries and aquaculture, FAO Fisheries and Aquaculture Technical Paper. 2017, 615. http://www.fao.org/3/a-i7677e.pdf

[14] OSPAR: OSPAR request on development of a common monitoring protocol for plastic particles in fish stomachs and selected shellfish on the basis of existing fish disease surveys. https://www.ices.dk/news-and-events/news-archive/news/Pages/Fish-stomachs-as-indicators-of-marine-litter.aspx

[15] M. Anthony Browne, T. Galloway Susan, R.C. Thompson: Spatial Patterns of Plastic Debris along Estuarine Shorelines. Environ. Sci. Technol., 2010, 44(9), 3404–3409.

[16] J.H. Dekiff, D. Remy, J. Klasmeier, E. Fries: Occurrence and spatial distribution of microplastics in sediments from Norderney. Environ. Pollut., 2014, 186, 248–256.

[17] C.P. Niemiec, R.M. Ryan: Autonomy, competence, and relatedness in the classroom: Applying self-determination theory to educational practice. Theory Res. Educ., 2009, 7(2), 133–144.

[18] P.D. Jones, K.R. Briffa, T.J. Osborn, E. Al: High-resolution palaeoclimatology of the last millennium: A review of current status and future prospects. The Holocene, 2009, 19(1), 3–49.

[19] J. Li, H. Liu, J.P. Chen: Microplastics in freshwater systems: A review on occurrence, environmental effects, and methods for microplastics detection. Water Res., 2018, 137, 362–374.

[20] J. Sun, X. Dai, Q. Wang, M.C.M. van Loosdrecht, B.-J. Ni: Microplastics in wastewater treatment plants: Detection, occurrence and removal. Water Res., 2019, 152, 21–37.

[21] Centre Ecotox News: Centre Suisse d'écotoxicologie appliquée : Les microplastiques dans l'environnement" 2015. https://www.ecotoxcentre.ch/media/25544/2015_mikroplastik_fr.pdf

[22] J.W. Desforges, M. Galbraith, P.S. Ross: Ingestion of Microplastics by Zooplankton in the Northeast Pacific Ocean. Arch. Env. Contam. Toxicol., 2015, 69(3), 320–330.

[23] D.W. Laist: Impacts of Marine Debris: Entanglement of Marine Life in Marine Debris Including a Comprehensive List of Species with Entanglement and Ingestion Records. 1997. DOI:10.1007/978-1-4613-8486-1_10.

[24] W. Wang, H. Gao, S. Jin, R. Li, G. Na: The ecotoxicological effects of microplastics on aquatic food web, from primary producer to human: A review. Ecotoxicol. Environ. Saf., 2019, 173, 110–117.

[25] S. Claessens, S.R. Ghosh, R. Mihet: Macro-Prudential Policies to Mitigate Financial System Vulnerabilities. Int. Monet. Fund, Journal of International Money and Finance 2013 39(155), 153–185; DOI:10.1016/j.jimonfin.2013.06.023

[26] T. Iwata: Biodegradable and Bio-Based Polymers: Future Prospects of Eco-Friendly Plastics. Angew. Chem. Int. Ed., 2015, 54, 3210–3215.

[27] L. Van Cauwenberghe, L. Devriese, F. Galgani, J. Robbens, C.R. Janssen: Microplastics in sediments: A review of techniques, occurrence and effects. Mar. Environ. Res., 2015, 111, 5–17.

[28] GESAMP: Sources, fate and effects of microplastics in the marine environment: A global assessment. International Marit. Organ. 4 Albert Embankment, London SE1 7SR, 2015.

[29] R.C. Thompson, S.H. Swan, C.J. Moore, F.S. Saal: Our plastic age," Philos. Trans. R. Soc. B Biol. Sci., 2009, 364, 1973–1976.

[30] M. Eriksen, L.C.M. Lebreton, H.S. Carson, M. Thiel, C.J. Moore, J.C. Borerro: Plastic Pollution in the World's Oceans: More than 5 Trillion Plastic Pieces Weighing over 250,000 Tons Afloat at Sea. PLoS ONE 2014, 9(12), e111913. https://doi.org/10.1371/journal.pone.0111913

[31] C.M. Rochman, E. Hoh, B.T. Hentschel, S. Kaye: Long-Term Field Measurement of Sorption of Organic Contaminants to Five Types of Plastic Pellets: Implications for Plastic Marine Debris. Environ. Sci. Technol., 2013, 47, 1646–1654.

[32] D.K.A. Barnes, F. Galgani, R.C. Thompson, M. Barlaz: Accumulation and fragmentation of plastic debris in global environments. Philos. Trans. R. Soc. London, B Biol. Sci., 2009, 364, 1985–1998.

[33] W. Sanchez, C. Bender, J.-M. Porcher: Wild gudgeons (*Gobio gobio*) from French rivers are contaminated by microplastics: Preliminary study and first evidence. Environ. Res., 2014, 128, 98–100.

[34] M.J. Hoffman, E. Hittinger: Inventory and transport of plastic debris in the Laurentian Great Lakes. Mar. Pollut. Bull., 2017, 115, 273–281.

[35] D. Lithner: Environmental and health hazards of chemicals in plastic polymers and products. 2011, DOI 978-91-85529-46-9; http://hdl.handle.net/2077/24978

[36] D. Lithner, A. Larsson, G. Dave: Environmental and health hazard ranking and assessment of plastic polymers based on chemical composition. Sci. Total Env., 2011, 409, 3309–3324.

[37] P. Morone, V.E. Tartiu, P. Falcone: Assessing the potential of biowaste for bioplastics production through social network analysis. J. Clean. Prod, 2015, 90, 43–54.

[38] T. Iwata: Biodegradable and Bio-Based Polymers: Future Prospects of Eco-Friendly Plastics. Angew. Chem. Int. Ed., 2015, 54, 3210–3215.

[39] K.G. Harding, T. Gounden, S. Pretorius: "Biodegradable" Plastics: A Myth of Marketing. Procedia Manuf., 2017, 7, 106–110.

INFLUENCE OF TYPICAL IONS AND NATURAL ORGANIC MATTER ON THE AGGREGATION OF MICROPLASTIC PARTICLES IN AQUEOUS PHASE

Runxi Liu, Hongtao Wang

[1] State Key Laboratory of Pollution Control and Resource Reuse, Key Laboratory of Yangtze River Water Environment, Ministry of Education, College of Environmental Science and Engineering, UNEP-TONGJI Institute of Environment for Sustainable Development, Tongji University, Siping Rd 1239, Shanghai 200092, P. R. China, hongtao@tongji.edu.cn

[2] Shanghai Institute of Pollution Control and Ecological Security, Shanghai 200092, P.R. China

Keywords: Aggregation, Inorganic Ions, Microplastics, Nano-plastics, NOM

Abstract

Micro-plastic contamination has become a global issue, and contamination of smaller-sized nano-plastics in water system has brought many concerns. Understanding the aggregation process of nano-plastics is a key step to analyze their fate and transport in aqueous phase. This article explores the effects of typical inorganic cations, anions, natural organic matter and pH on the aggregation of polystyrene nanoparticles (PS NPs) in aqueous phase. The result shows that PS NPs remained substantially stable in low ionic strength (NaCl: 1-200 mM or $CaCl_2$: 0.1-5 mM), whereas significant aggregation occurred in high ionic strength (NaCl: 400 mM or $CaCl_2$: 10 mM). In addition, under the same ionic strength conditions, the particles remained more stable in Na_2SO_4 solution rather than in NaCl. The addition of natural organic matter had no significant effect on the aggregation of PS NPs at 150 mM NaCl, 0.1-1 mM $CaCl_2$, or 1/3 - 50/3 mM Na_2SO_4. The aggregation of PS NPs was boosted at 5 mM $CaCl_2$, while it was inhibited at 200 mM NaCl or 200/3 - 400/3 mM Na_2SO_4. When pH was close to the isoelectric point of PS, the condensation effect was significantly enhanced and the aggregation of PS NPs promoted. Based on the influence of several different factors, the aggregation behavior of PS NPs can be fitted and predicted by using DLVO theory.

1 Introduction

Microplastics (MPs) are synthetic polymer particles of a certain size, normally with diameters from 1 μm to 5 mm. According to the formation process, there are primary and secondary plastics. Primary plastics are small plastic particles contained in products or raw materials, such as cosmetics, cleaning products [1, 2] and plastic resin powder [3, 4], etc. Secondary plastics are decomposed products of bulky plastics, such as fiber shedding caused by washing clothes [5] or plastics wearing outside [6]. Plastic can accumulate in the aquatic system [7], and existing wastewater treatment process is difficult to remove small plastic particles [1]. A growing number of studies have shown that synthetic polymers, especially MPs, are ubiquitous in the environment and various ecosystems, including sediments, soils and surface layers in marine and freshwater systems [8, 9]. Cyclic patterns indicate that there are accumulation areas in all five Subtropical Ocean circulations [10]. With the migration of natural conditions such as wind and ocean currents, the pollution of MPs has become a common topic over the world.

Few studies on the fate and transport of nano-plastics (NPs) had been carried out until recent years [11]. Factors in the environment, such as pH, inorganic ions, natural organic matter (NOM), suspended clay particles and bacteria, are key issues to NPs aggregation. For now, the mechanism of these factors is still unclear. Some scholars believe that characteristics of NPs are similar to nanoparticles used in engineering [12]. Cai et al. [13] explored the influence of inorganic ions (Na^+, Ca^{2+}, Fe^{3+}) and natural organic matter (SRHA, SRFA) on NPs aggregation through dynamic light scattering (DLS), finding out that higher valence cations can promote the aggregation.

Polystyrene (PS) microspheres, as a typical representative of MPs, have been widely used as plasticizers, antioxidants and flame retardants [14, 15]. Researchers have studied the colloidal stability of PS microspheres in aquatic environment under different conditions [13, 16], indicating that the aqueous solution conditions (i.e., pH, ionic strength and humic acid content) can significantly affect the surface charge of PS microspheres in solutions or electrostatic bilayers such as stereo repulsion [17, 18]. However, humic acid (HA) can enhance the stability of MPs by electrostatic repulsion or steric hindrance [19].

This article aims to investigate the influence of aqueous solution conditions (usually inorganic ions, NOM and pH) on NPs aggregation. It is expected to provide information for the fate and transfer of NPs in aqueous phase.

2 Materials and Methods

2.1 Materials

The yellow-green fluorescent polystyrene nanoparticles (PS NPs) with a diameter of 0.1 μm were purchased from Thermo-Fisher Corp., USA and used as received without surface modification. The PS NPs stock suspension was hydrophilic and electrostatically stabilized (according to the manufacturer) with a particle concentration of 10 g/L and a density of 1.05 g/cm^3. All samples were dispersed by ultrasound before use. NaCl (AR), $CaCl_2$ (AR), and Na_2SO_4 (AR) salts were purchased from Sinopharm Chemical Reagent Co., Ltd and used as received without further purification. Humic acid (#53680-10G, Sigma-Aldrich) was used as natural organic matter. 40 mg/L humic acid stock solutions were prepared by dissolving 20.0 mg powder in 500 mL Milli-Q water with pH 11 and 24 h stirring until no solid left. After filtering through 0.25 μm membranes, the stock solutions were stored in 4 °C fridges. Determined by a TOC-meter (Shimadzu, Japan), the total organic carbon content of 10 mg/L humic acid solutions was 4.5 mg/L.

2.2 Methods

2.2.1 Measurement of zeta potential and morphology

Zetasizer Nano ZS90 (Malvern Instruments, UK) was used to measure zeta potential of PS NPs performing 10-100 times for each sequence at room temperature (25 °C). The average of 3 sequences for each sample was taken as the results.

Floc samples sprayed with gold were observed by Phenom G2 Pro SEM (Phenom China) after 60 min mixing of PS NPs and solutions. With proper dilution, the mixture samples were dried and fixed on a silicon chip by electric tape.

2.2.2 Size growth measurement

The number-weighted averaged hydrodynamic size of PS NPs was determined by time-resolved DLS as a function of time. Experiments were performed on Zetasizer nano ZS90 (Malvern UK) with 1 min DLS measurement for 60 times, operating with a He-Ne laser at a wavelength of 633 nm and at a scattering angle of 90°.

pH was set as 6.3 and 9.18 for different experiments. According to the ionic strength (IS) in the natural environment [20], NaCl solutions were prepared at 1, 50, 200, 400 mmol/L, $CaCl_2$ at 0.3, 3, 15, 30 mmol/L and Na_2SO_4 at 1, 50, 200, 400 mmol/L. The concentration of PS NPs in each experiment was 50 mg/L.

In this study, the condensation rate was defined as the slope of particle size growth ($\Delta D/\Delta t$), where ΔD represents the increase of NPs' hydraulic diameter (nm) in a certain period of time Δt (min). Besides, the energy barrier based on DLVO theory was calculated in order to better understand the aggregation mechanisms.

3 Results and Discussion

3.1 Characteristic of PS NPs

The PS NPs sample had a narrow distribution of sizes with an average size of 104.8 nm (Figure 1 a). The zeta potential of PS NPs was negative from pH 3 to 10 (Figure 1 b), and the absolute value increased with the pH (exception pH 3, where PS NPs could be damaged in strong acid), indicating a rather stable status.

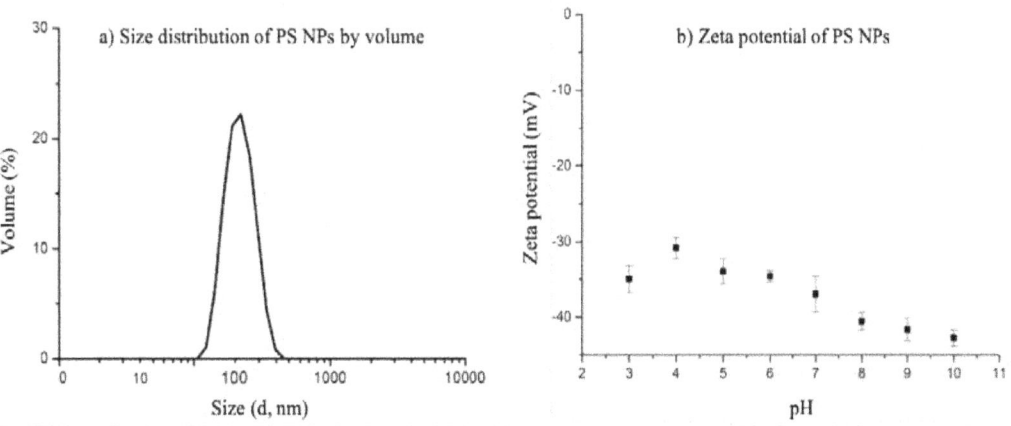

Figure 1: (a) Size distribution (DLS, pH 6.3, 10 mg/L PS) and (b) Zeta potential (pH 3-10, 10 mg/L PS) of original PS NPs samples

3.2 Influence of NaCl and NOM on PS NPs aggregation

As shown in Figure 1, the PS NPs remained stable in NaCl solutions at ionic strength from 1 to 50 mM, no matter in the absence or presence of NOM and did not aggregate until at 200 mM IS with the diameter increasing to ~200 nm. When IS of NaCl solution was 400 mM, the PS NPs showed obvious aggregation with ~300 nm growth in diameter. However, NOM inhibited PS NPs' size growth at 200 and 400 mM IS, respectively, also illustrated by growth rate of particle size (Table 1). The change of hydrodynamic sizes of PS NPs (10 mg/L) in Milli-Q water was measured as control.

The results indicated that aggregation of PS NPs in NaCl was boosted by the increase of solutions' IS (Figure 2 a) and inhibited by NOM for the increase of electrostatic and steric stabilization (Figure 2 b).

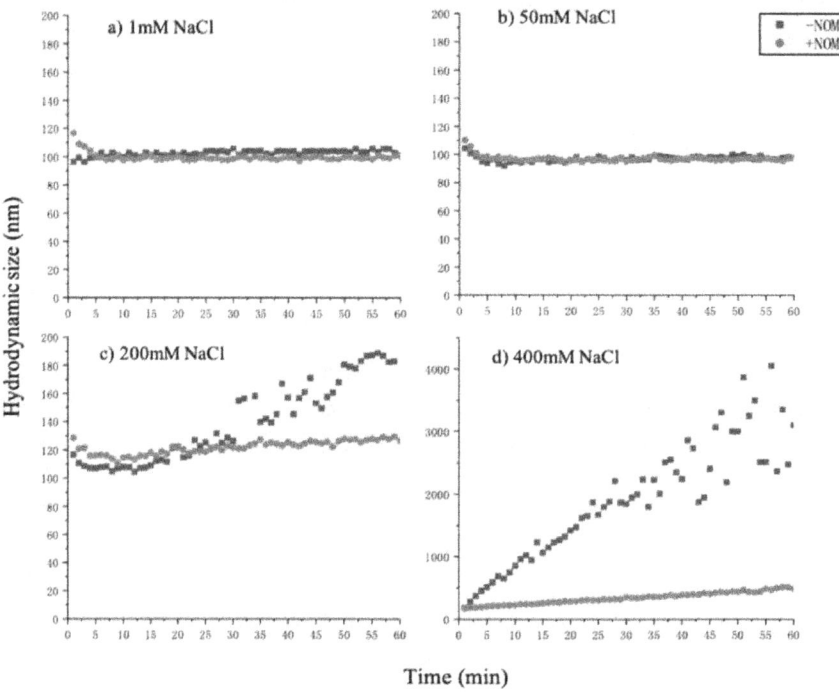

Figure 2: Change of hydrodynamic size of PS NPs in (a) 1 mM, (b) 50 mM, (c) 200 mM and (d) 400mM NaCl solutions

Table 1: Hydrodynamic size growth rate of PS NPs in different NaCl solutions

Ionic strength (mM)	1	50	200	400
ΔD/Δt (nm/min)	0.04	0.08	1.60	49.98
ΔD/Δt (+NOM, nm/min)	-0.04	-0.05	0.22	5.35

Note: NOM concentration = 10 mg/L

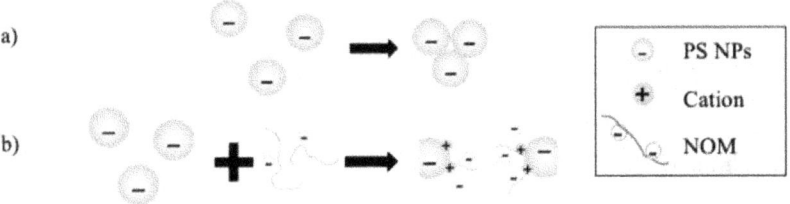

Figure 3: Illustration of PS NPs in NaCl solutions (a) without NOM and (b) with NOM

3.3 Influence of CaCl$_2$ and NOM on PS NPs aggregation

From Figure 4, no aggregation of PS NPs was observed until the ionic strength increased to 30 mM with the diameter reaching ~200 nm. The addition of NOM at ionic strength 0.3 and 3 mM had limited influence on PS NPs' aggregation. However, NOM promoted the hydrodynamic size of PS NPs at an ionic strength of 15 or 30 mM and double the growth rate even at ionic strength 30 mM (Table 2).

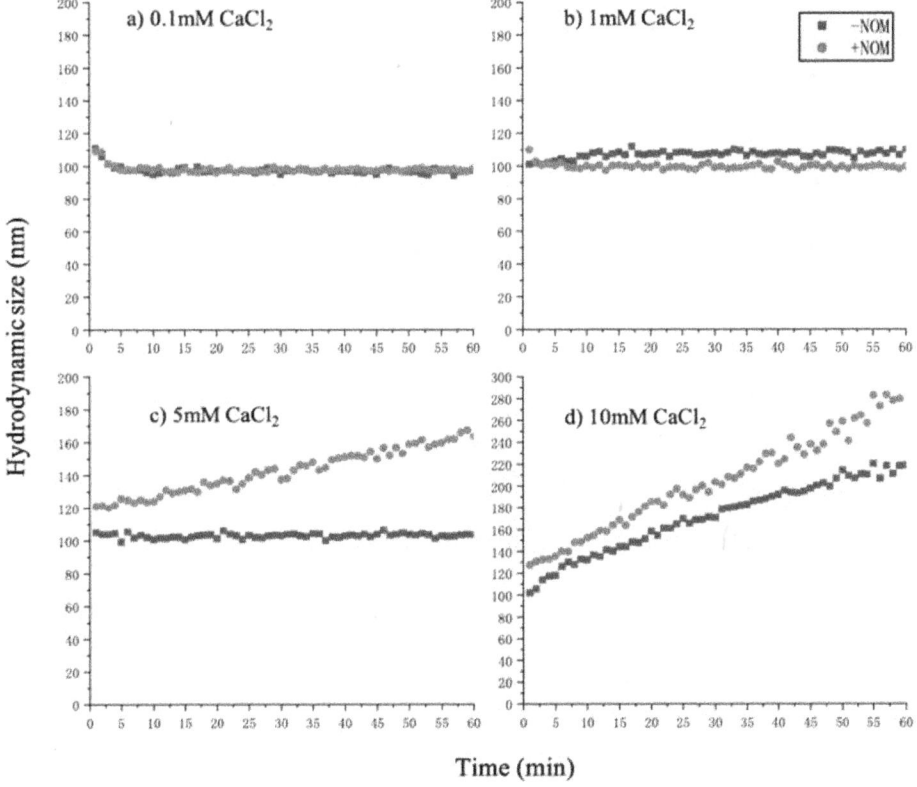

Figure 4: Change of hydrodynamic size of PS NPs in (a) 0.1 mM, (b) 1 mM, (c) 5mM and (d) 10 mM CaCl$_2$ solutions

Table 2: Hydrodynamic size growth rate of PS NPs in different CaCl$_2$ solutions

Ionic strength (mM)	0.3	3	15	30
ΔD/Δt (nm/min)	-0.06	0.07	0.01	1.85
ΔD/Δt (+NOM, nm/min)	-0.04	-0.03	0.76	2.61

Note: NOM concentration = 10 mg/L

By means of energy barrier calculation based on DLVO theory, the reason for NOM boosting aggregation in CaCl$_2$ at IS 15 and 30 mM could be due to the presence of NOM building a network structure through capturing effect of electrostatic attraction along with PS NPs covered with Ca^{2+} (Figure 6 a), and therefore, lowering the energy barrier between PS NPs, corresponding to the calculation result (Figure 5). From the SEM image (Figure 7), it is clear that PS NPs in CaCl$_2$ solution at IS 30 mM with NOM formed a denser network structure.

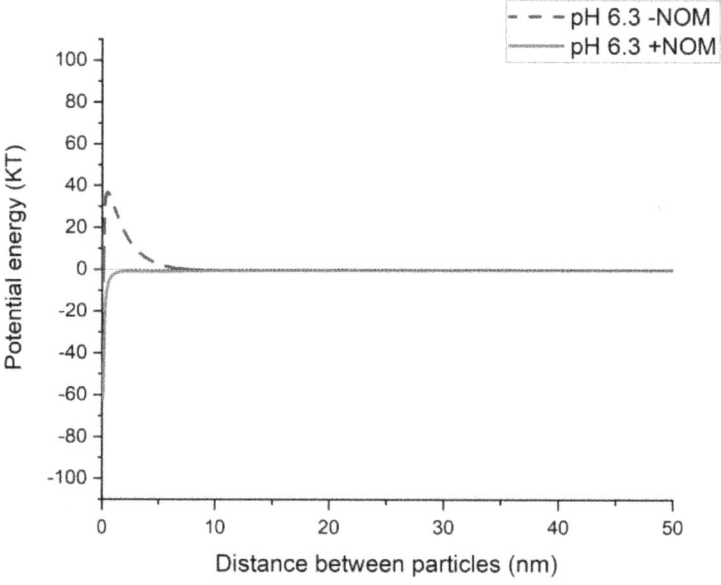

Figure 5: Energy barrier calculation of PS NPs in 10 mM CaCl$_2$ solutions

It is noteworthy that at lower IS (0.3 and 3 mM), the aggregation still was not observed in the presence of NOM. Possible explanation is that the concentration of cation Ca^{2+} was too low to modify surface charge characteristics of PS NPs (Figure 6 b).

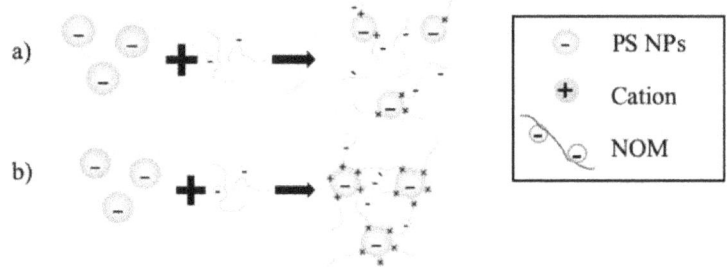

Figure 6: Illustration of PS NPs in (a) 5, 10 mM and (b) 0.1, 1 mM CaCl$_2$ solutions with NOM

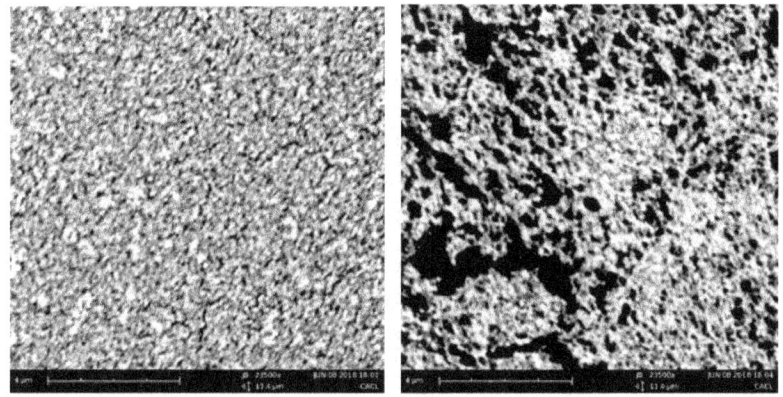

Figure 7: SEM images of PS NPs in 10mM CaCl$_2$ solutions with and without NOM
(*left*: 10 mM CaCl$_2$ + NOM; *right*: 10 mM CaCl$_2$)

3.4 Influence of Na$_2$SO$_4$ and NOM on PS NPs aggregation

Similar to the results of NaCl, the hydrodynamic size of PS NPs barely changed at ionic strength from 1 to 50 mM and increased to ~350 nm and ~1000 nm at 200 mM and 400 mM ionic strength, respectively (Figure 8). The presence of NOM also caused inhibition of aggregation at IS 200 and 400 mM, while hardly influenced the particle size at lower IS (Table 3).

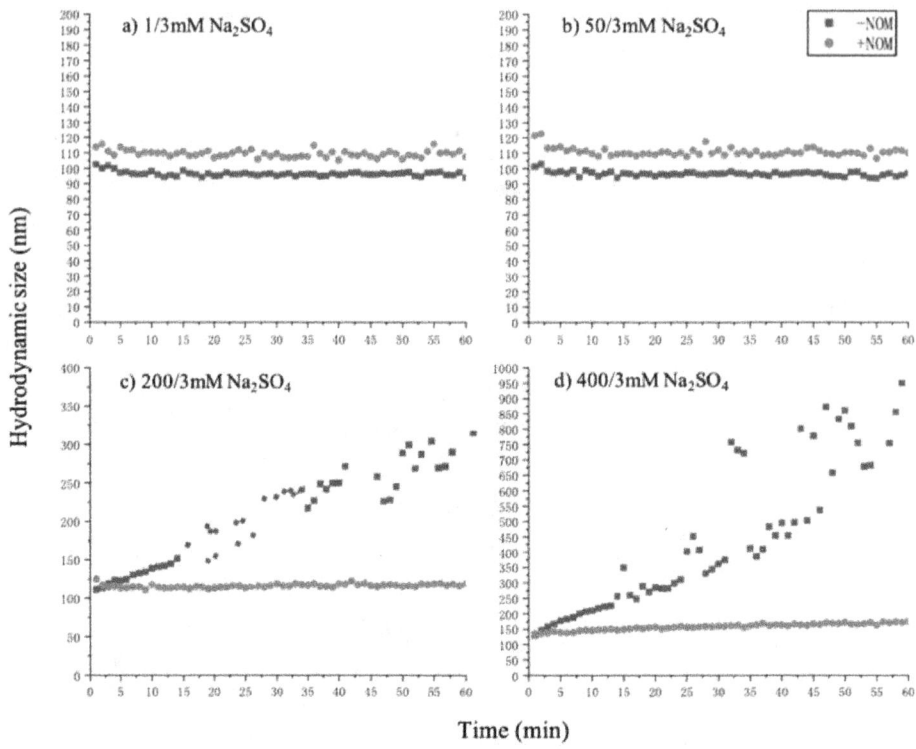

Figure 8: Change of hydrodynamic size of PS NPs in (a) 1/3 mM, (b) 50/3 mM,(c) 200/3 mM and (d) 400/3 mM Na_2SO_4 solutions

Table 3: Hydrodynamic size growth rate of PS NPs in different Na_2SO_4 solutions

Ionic strength (mM)	1	50	200	400
ΔD/Δt (nm/min)	0.02	0.05	3.45	14.57
ΔD/Δt (+NOM, nm/min)	-0.03	-0.04	0.06	0.57

Note: NOM concentration = 10 mg/L

3.5 Influence of pH on PS NPs aggregation

Compared with pH 9.18, PS NPs in $CaCl_2$ solution (IS = 30 mM) at pH 6.3 was observed to have a significant aggregation with nearly doubled particle size. At the same time, based on results from section 3.3, the aggregation was inhibited at pH 9.18 (Figure 9) with even smaller zeta potential (Table 4) and higher energy barrier (Figure 10).

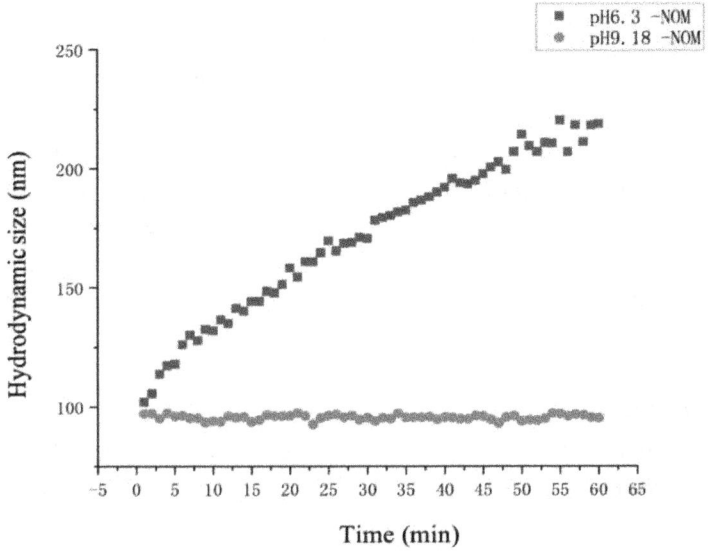

Figure 9: Change of hydrodynamic size of PS NPs in 10 mM CaCl$_2$ solutions at pH 6.3 and 9.18

Table 4: Zeta potential of PS NPs in 10 mM CaCl$_2$ at pH 6.3 and 9.18

pH	Zeta potential (mV)	Status
6.3	-21.34	Slightly unstable
9.18	-32.26	Stable

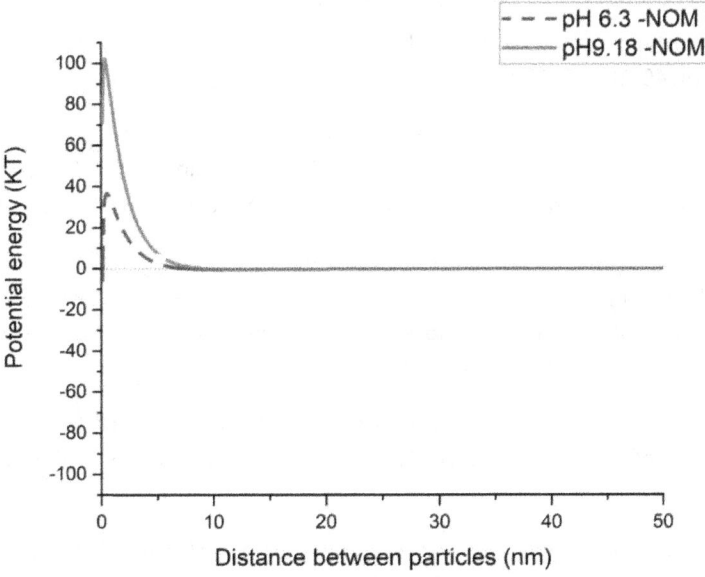

Figure 10: Energy barrier calculation of PS NPs in 10 mM CaCl$_2$ solutions at pH 6.3 and 9.18

4 Conclusions

Three key factors, namely pH, ionic strength and NOM have different influences on the aggregation of PS NPs. The hydrodynamic size of PS NPs increases with ionic strength; and Ca(II) cation with higher valence, has stronger influence than Na(I) at the same IS for its larger hydration radius. There is also a positive correlation relationship between pH and stability of PS NPs. However, influence of NOM is more complicated. In Ca(II) solutions, the presence of NOM boosted aggregation at IS 15 and 30 mM, benefiting from the large molecular weight and network structure of NOM, while in Na(I) solutions NOM showed a negative effect for steric hindrance brought by NOM, which was stronger than electrostatic attraction with Na(I) having a smaller hydration radius.

5 Acknowledgements

The authors would like to thank DAAD and the Exceed Swindon project for supporting their participation at this workshop. They also sincerely thank Prof. Müfit Bahadir for editing the manuscript.

6 References

[1] Carr, S.A., Liu J., Tesoro A.G.: Transport and fate of microplastic particles in wastewater treatment plants. Water Research, 2016, 91, 174-182.

[2] Duis, K., Coors, A.: Microplastics in the aquatic and terrestrial environment: sources (with a specific focus on personal care products), fate and effects. Environmental Sciences Europe, 2016, 28(1), 2.

[3] Zbyszewski, M., Corcoran, P.L., Hockin, A.: Comparison of the distribution and degradation of plastic debris along shorelines of the Great Lakes, North America. Journal of Great Lakes Research, 2014, 40(2), 288-299.

[4] Chen, C.S., Anaya, J.M., Zhang, S., Spurgin, J., Chuang, C.Y., Xu, C., Miao, A.J., Eric, Y.T., Schwehr, K.A., Jiang, Y.L., Quigg, A., Santschi, P.H., Chin, W.C.: Effects of Engineered Nanoparticles on the Assembly of Exopolymeric Substances from Phytoplankton. Plos One, 2011, 6(7), e21865.

[5] Browne, M.A., Crump, P., Niven, S.J., Teuten, E., Tonkin, A., Galloway, T., Thompson, R.: Accumulation of microplastic on shorelines worldwide: sources and sinks. Environmental Science & Technology, 2011, 45(21), 9175-9179.

[6] Tosin, M., Weber, M., Siotto, M., Lott, C., Degli, I.F.: Laboratory Test Methods to Determine the Degradation of Plastics in Marine Environmental Conditions. Frontiers in Microbiology, 2012, 3, 225.

[7] Alimi, O.S., Farner, B.J., Hernandez, L.M., Tufenkji, N.: Microplastics and Nanoplastics in Aquatic Environments: Aggregation, Deposition, and Enhanced Contaminant Transport. Environmental Science & Technology, 2018, 52(4), 1704-1724.

[8] Li, W.C., Tse, H.F., Fok, L.: Plastic waste in the marine environment: A review of sources, occurrence and effects. Science of The Total Environment, 2016, 566-567, 333-349.

[9] Chae, Y., An, Y.J.: Effects of micro- and nanoplastics on aquatic ecosystems: Current research trends and perspectives. Marine Pollution Bulletin, 2017, 124(2), 624-632.

[10] Lebreton, C.M., Greer, S.D., Borrero, J.C.: Numerical modelling of floating debris in the world's oceans. Marine Pollution Bulletin, 2012, 64(3), 653-661.

[11] Eerkesmedrano, D., Thompson, R.C., Aldridge, D.C.: Microplastics in freshwater systems: A review of the emerging threats, identification of knowledge gaps and prioritisation of research needs. Water Research, 2015, 75(3), 63-82.

[12] Long, M., Paul-Pont, I., Hégaret, H., Moriceau, B., Lambert, C., Huvet, A., Soudant, P.: Interactions between polystyrene microplastics and marine phytoplankton lead to species-specific hetero-aggregation. Environmental Pollution, 2017, 228, 454-463.

[13] Cai, L., Hu, L., Shi, H., Ye, J.W., Zhang, Y.F., Kim, H.J.: Effects of inorganic ions and natural organic matter on the aggregation of nanoplastics. Chemosphere, 2018, 197, 142-151.

[14] Della, T.C., Bergami, E., Salvati, A., Faleri, C., Cirino, P., Dawson, K.A., Corsi, I.: Accumulation and embryotoxicity of polystyrene nanoparticles at early stage of development of sea urchin embryos *Paracentrotus lividus*. Environmental Science & Technology, 2014, 48(20), 12302-11.

[15] Tan, L., Tan, B.: Hypercrosslinked porous polymer materials: design, synthesis, and applications. Chemical Society Reviews, 2017, 46(11), 3322.

[16] Jódarreyes, A.B., Ortegavinuesa, J.L., Martínrodríguez, A.: Electrokinetic behavior and colloidal stability of polystyrene latex coated with ionic surfactants. Journal of Colloid & Interface Science, 2006, 297(1), 170-181.

[17] Drechsler, A., Grundke, K.: The influence of electrolyte ions on the interaction forces between polystyrene surfaces. Colloids & Surfaces A Physicochemical & Engineering Aspects, 2005, 264(1-3), 157-165.

[18] Phenrat, T., Song, J.E., Cisneros, C.M., Schoenfelder, D.P., Tilton, R.D., Lowry, G.V.: Estimating attachment of nano- and submicrometer-particles coated with organic macromolecules in porous media: development of an empirical model. Environmental Science & Technology, 2010, 44(12), 4531-8.

[19] Chen, W., Ouyang, Z.Y., Qian, C., Yu, H.Q.: Induced structural changes of humic acid by exposure of polystyrene microplastics: A spectroscopic insight. Environmental Pollution, 2018, 233, 1-7.

[20] Tang, S.C.N., Birnhack, L., Cohen, Y., Lahav, O.: Selective separation of divalent ions from seawater using an integrated ion-exchange/nanofiltration approach. Chemical Engineering and Processing - Process Intensification, 2018, 126, 8-15.

MICROPLASTICS IN HUMAN CONSUMPTION: TABLE SALT CONTAMINATED WITH MICROPLASTICS

E.V. Ramasamy, S.N. Sruthi, A.K. Harit, N. Babu

School of Environmental Sciences, Mahatma Gandhi University, Kottayam-686 560, Kerala, India;
evramasamy@mgu.ac.in

Keywords: Table salt, human consumption, microplastics, Raman spectroscopy, polymers

Abstract

Plastic debris of less than 5 mm size, known as Microplastics (MPs), have been found in aquatic and terrestrial environment, and it has become a global concern. In India, reports on MPs are scarce and very limited as far as their presence in items being consumed by human are concerned. There exists only a single report on the presence of MPs in table salt from India, and no report exists particularly on the national and local brands of salts available in the southern part of India. Present study aimed to investigate six brands of table salts available in Indian market for the presence of MPs. Stereomicroscopy, fluorescent microscopy and Raman spectroscopy were used in the process of identification of MPs and to study the polymer content of the same. The findings indicate the presence of MPs in all brands of salts examined in this study. To the best of our knowledge, the current study is the first report on the presence of MPs in table salt samples from Kerala – the state located in the southernmost tip of the country.

1 Introduction

Plastic pollution is a growing concern at global level. Ever increasing production of plastics and unscientific disposal methods has increased plastic pollution in aquatic and terrestrial environment throughout the world. Plastic wastes, unlike other human-generated organic garbage, persist in the environment for years, polluting the environment and affecting the organisms enormously. About 80% of plastic debris in oceans originate from land and are transported through the network of water bodies including lakes, streams, rivers, estuaries and wetlands [1]. Plastic debris in the aquatic environment break down due to the sun light or the mechanical action of wind, waves or biological means into tiny particles of microscopic sizes referred to as 'microplastics' (MPs) often referred to particles of < 5 mm in size. The impact of MPs pollution on the environment and biota is being researched globally. It is due to their small size, they are bioavailable to the organisms of the food-web in aquatic environment. Large surface-to-volume ratio and chemical composition of the MPs aid in accumulation of waterborne contaminants including Persistent, Bio accumulative and Toxic (PBT) compounds [2].

Studies on MPs began in early 2000s; more reports are available on MPs in the marine environment [2, 3], while information on freshwater and estuarine systems are scarce [4]. When the marine and freshwater systems are contaminated heavily with MPs, it is obvious that they will end up in the products obtained from these water bodies. Hence, studies on the contamination of MPs in biotic products from ocean/sea, such as fish, clams, mussels, etc., and in abiotic products, such as sea salts are being carried out in many parts of the world. MPs have been reported in marine organisms, including plankton, fish, and mammals [5, 6], also in commercially available food items, such as honey, beer, salt, and drinking water [7-11]. However, number of publications on MPs contamination in sea salt, which is under human consumption, is relatively less. For instance, global attention on the presence of MPs in table salt began only in 2015 [12], and currently there are only 12 research publications available on table salt contaminated with MPs, and only one is from India. The objective of this study is to investigate the occurrence of MPs in several brands of commercially available table salts in India.

2 Material and Methods

2.1 Sampling and analyses

In the present study, six brands of table salts were collected from the supermarkets/shops in Kerala, India. Out of these, two were from internationally available brands, two were popular throughout the country and two local brands of salt available in Kerala. The national and international brands of salt were in powder form, packed in Polyethylene (PE) packets, whereas the local brand of salt was in crystal form, also packed in PE packets. The national brands of salt were named as NS 1 and NS 2; the international brands were denoted as IS 1, IS2, and the local brand was named as LS1 and LS2.

Standard protocol as prescribed by National Oceanic and Atmospheric Administration (NOAA) for extraction of MPs including floatation, filtration, and separation was followed in this work. Salt samples of 270 g were taken in a glass beaker and dissolved in 1 L double distilled water. After the completion of dissolution, these samples (whole 1 L) were centrifuged at 1,900 rpm for 15 min. The supernatant after centrifugation were immediately filtered through vacuum filtration using glass fibre filter GF/A (< 0.45 µm pore size). After the filtration, the filter papers with MPs were placed into a pre-cleaned, pre-dried glass petri dish with glass cover and air dried at ambient temperature.

All equipment used in this study was rinsed thrice with ultrapure water [1]. The extraction process of MPs was performed in a clean fume hood, and the work surface was wiped with 70% ethyl alcohol. Double distilled water was used during extraction and filtration stages. Ultra-pure water was used as laboratory blanks in order to test for cross contamination of plastic materials during extraction and filtration processes.

After air drying, the filter papers were observed under stereomicroscope (Carl Zeiss, Stemi 508), for the identification of MPs, their shape and size. Fluorescent microscope (Olympus BX43 along with the light source 130 W U-HGLGPS light guide-coupled illumination system and Olympus DP27 digital color camera) was also used to identify the MPs particles having fluorescent properties and also to take their images. Polymer types of MPs were characterized using a micro Raman spectrometer (WITee Alpha 300RA, Germany) with a 532 nm excitation laser (grating of 600 groves/mm using 50X objective) with integration time of 10 s. The resulting spectra were compared with the KnowItAll Raman spectral library for polymers provided by BioRad.

2.2 Blank test for quality assurance

In order to substantiate the presence and distribution of MPs from the commercial salt samples, analytical reagent grade sodium chloride (NaCl, Emparta, Merk) and distilled water (the materials used for MPs extraction) were also analyzed separately. Artificial saline water was prepared by mixing up of 270 g NaCl in 1 L distilled water. Separation and vacuum filtration was done as per the procedure mentioned above. After the filtration, the glass fibre filter paper was analyzed with stereomicroscopy, fluorescent microscopy and Raman spectroscopy. No polymers were observed from the procedural blank. Besides that, NaCl and distilled water used in the study were free from the MPs contamination. Hence, the laboratory condition can be considered as clean.

3 Results and Discussion

The findings indicate the presence of MPs in all the six brands of salt examined in this study. Higher numbers of MPs were recorded with local brand salt, while international brands had the minimum number. Fibers were the predominant type of MPs observed with the salts studied. No polymers were noted in the laboratory blanks operated in this study.

3.1 Number, shape and size of MPs

The number of MPs particles varied between different brands of salt. A total of 264 MPs particles were extracted from all brands of salts put together. Out of this, a maximum of 118 MPs were found with local brand, followed by 90 particles in national and 55 number of MPs with the international brand (Figure 1). Most of the extracted MPs were in the shape of fibers and fragments of different sizes, while sheets were the least type of MPs (Figure 2).

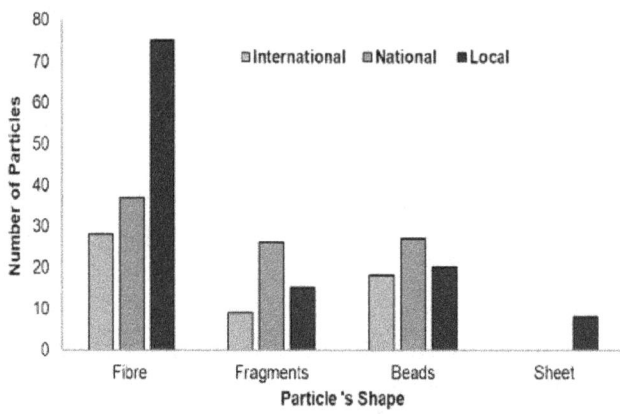

Figure 1: Number of particles in each type of MPs in different brands of salt

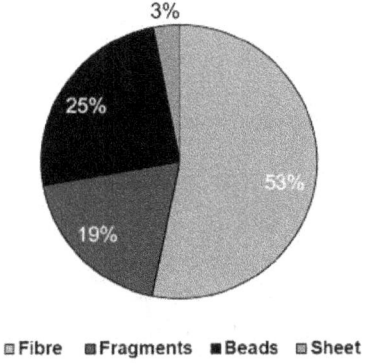

◻ Fibre ■ Fragments ■ Beads ◻ Sheet

Figure 2: Proportion of different types of MPs in Indian salt

Fibers were the predominantly observed MPs particles from all the table salt having different colors (transparent, blue, green, red, white, etc.). The images of MPs of stereo and fluorescent microscopes are provided in Figure 3. As far as the sizes of MPs are concerned, 30% of MPs fall under the size range of 500 to 1000 μm, while 11% of MPs belong to < 100 μm. The local brand of salt had the maximum MPs of smallest size of < 100 μm (which constitutes 76 %) as well as largest size of > 1000 μm. In summary, the local brands of salt had the maximum number of MPs, and the smallest as well as largest sized MPs were also noted more in this category.

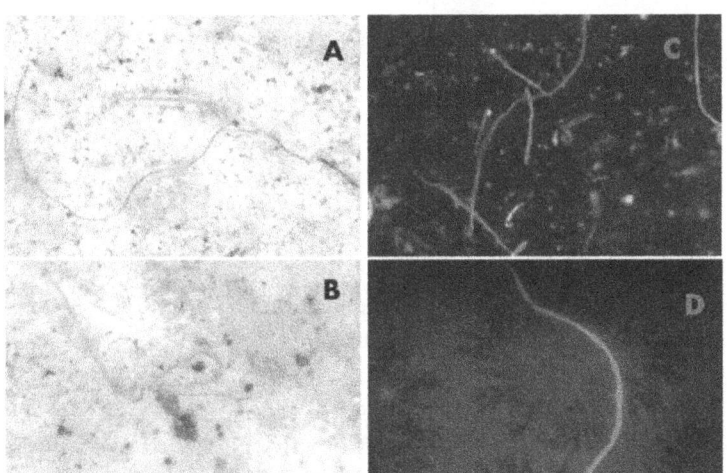

Figure 3: Stereo microscopic (A, B) and fluorescent microscopic (C, D) images of MPs

3.2 Polymer component of MPs identified with Raman Spectroscopy
The MPs extracted from salt were examined with Confocal Raman spectroscopy and revealed the presence of six types of polymers: Polystyrene (PS), Polyamide (PA), Polyethylene terephthalate (PET), Polypropylene (PP), Polyethylene (PE) and Polyvinyl Chloride (PVC). Polystyrene is the predominantly found polymer among the MPs, while PVC is the least observed polymer type.

International brand of salt had four type of polymers (PS > PET > PE > PVC), while rest of the brands had five type of polymers (Figure 4).

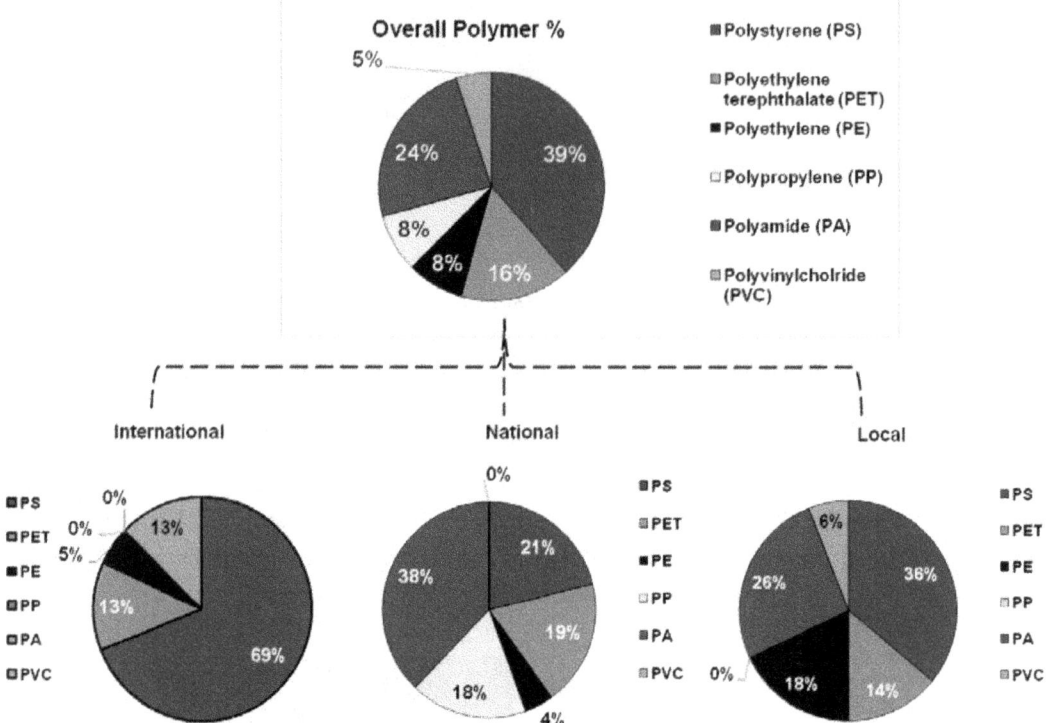

Figure 4: Polymer components of MPs extracted from different brands of salt

4 Discussion

This study revealed that the Indian sea salt is contaminated with MPs particles of varying size, shape and polymer composition. This confirms the hypothesis that the sea water has good number of MPs [13], thus the salt derived from sea water definitely possesses MPs. Besides sea water, contamination of salt during production and packaging through airborn MPs is also possible [12].

The average concentration of MPs in the sea salt brands analyzed in this study is 34 particles/kg salt. This is slightly lower than [14] reported for Indian salt, but it is more close to the observations of [15], reported for Turkish sea salt and that of Spanish ones [16]. However, it is lower than the MPs particle reported for China by [12] and higher MPs reported by [17].

The average daily intake of salts for humans has been reported by [18] as 5 g/d (according to WHO guidelines). The findings of the present study reveal that 34 MPs/kg of Indian salt has been noted, thus as per WHO recommendation the annual intake of MPs through Indian salt consumption becomes 62 MPs/capita.yr.

The MPs presence in the sea salt becomes hazardous due to the polymer components of the MPs and the additives, dyes, plasticizers, etc. Besides these, other possible toxic effects of MPs over the organisms include release of POPs (persistent organic compounds) and other chemicals, and the microbial biofilms adsorbed on MPs surface [16].

The toxicological risks of the consumption of MPs on humans are yet to be established; hence it is not possible to attribute MPs for any particular kind of sickness or disease. Thus, the toxicological aspects of MPs on humans need to be studied.

5 Conclusions

The present study reveals the occurrence of MPs in all the brands of sea salt commercially available in the state of Kerala, India. Even though the number of MPs found in the Indian salt is lower than the MPs concentration reported in the rest of the world, the harmful impacts on human health cannot be ignored. Necessary steps need to be initiated to avoid airborn sources of MPs contamination in the production and packaging units of salt industries. Proper plastic waste management systems are to be established in order to reduce the inflow of plastic debris into the water bodies so that the future generation could be protected from the issue of MPs.

6 Acknowledgements

Authors gratefully acknowledge DST-PURSE (Phase II), DST–SAIF, DST-N-PDF fellowship (PDF/2017/002637) for the financial support and instrumentation facility extended to this work. We also acknowledge the support of Exceed-Swindon for workshop participation.

7 References

[1] Sruthy, S., Ramasamy, E.V.: Microplastic pollution in Vembanad Lake, Kerala, India: The first report of microplastics in lake and estuarine sediments in India, Environmental Pollution 2017, 222, 315-322.

[2] Cole, M., Lindeque, P., Halsband, C., Galloway, T.S.: Microplastics as contaminants in the marine environment: a review. Marine Pollution Bulletin 2011, 62(12), 2588-2597.

[3] Thompson, R.C., Olsen, Y., Mitchell, R.P., Davis, A., Rowland, S.J., John, A. W.,..., Russell, A.E.: Lost at sea: where is all the plastic? Science 2004, 304(5672), 838-838.

[4] Free, C.M., Jensen, O.P., Mason, S.A., Eriksen, M., Williamson, N.J., Boldgiv, B.: High-levels of microplastic pollution in a large, remote, mountain lake. Marine Pollution Bulletin 2014, 85(1), 156-163.

[5] Desforges, J.P.W., Galbraith, M., Ross, P.S.: Ingestion of microplastics by zooplankton in the Northeast Pacific Ocean, Archives of Environmental Contamination and Toxicology, 2015, 69(3), 320-330.

[6] Wiesheu, A.C., Anger, P.M., Baumann, T., Niessner, R., Ivleva, N.P.: Raman microspectroscopic analysis of fibers in beverages. Analytical Methods 2016, 8(28), 5722-5725.

[7] Mintenig, S.M., Löder, M.G.J., Primpke, S., Gerdts, G.: Low numbers of microplastics detected in drinking water from ground water sources. Science of The Total Environment, 2019, 648, 631-635.

[8] Pivokonsky, M., Cermakova, L., Novotna, K., Peer, P., Cajthaml, T., Janda, V.: Occurrence of microplastics in raw and treated drinking water. Science of The Total Environment, 2018, 643, 1644-1651.

[9] Welle, F., Franz, R.: Microplastic in bottled natural mineral water – Literature review and considerations on exposure and risk assessment. Food Additives & Contaminants: Part A 2018, 35(12), 2482-2492.

[10] Schymanski, D., Goldbeck, C., Humpf, H.U., Fürst, P.: Analysis of microplastics in water by micro-Raman spectroscopy: release of plastic particles from different packaging into mineral water. Water Research, 2018, 129, 154-162.

[11] Kosuth, M., Mason, S.A., Wattenberg, E.V.: Anthropogenic contamination of tap water, beer, and sea salt. PloS One, 2018, 13(4), e0194970.

[12] Yang, D., Shi, H., Li, L., Li, J., Jabeen, K., Kolandhasamy, P.: Microplastic pollution in table salts from China. Environmental Science & Technology, 2015, 49(22), 13622-13627.

[13] Isobe, A., Uchiyama-Matsumoto, K., Uchida, K., Tokai, T.: Microplastics in the Southern Ocean. Marine Pollution Bulletin, 2017, 114(1), 623-626.

[14] Seth, C.K., Shriwastav, A.: Contamination of Indian sea salts with microplastics and a potential prevention strategy. Environmental Science and Pollution Research, 2018, 25(30), 30122-30131.

[15] Gündoğdu, S.: Contamination of table salts from Turkey with microplastics. Food Additives & Contaminants: Part A, 2018, 35(5), 1006-1014.

[16] Iñiguez, M.E., Conesa, J.A., Fullana, A.: Microplastics in Spanish table salt. Scientific Reports, 2017, 7(1), 8620.

[17] Karami, A., Golieskardi, A., Choo, C.K., Larat, V., Galloway, T.S., Salamatinia, B.: The presence of microplastics in commercial salts from different countries. Scientific Reports, 2017, 7, 46173.

[18] Domnariu, C.D., Cucu, A., Furtunescu, F.L.: World Health Organization guidelines on salt intake in adults and children. Acta Medica Transilvanica, 2013, 18(1), 166.

MICROPLASTICS REMOVAL FROM MUNICIPAL WASTEWATER BY AN ELECTROCOAGULATION SYSTEM

G.H. Bracher[1], C. Graepin[1], D. Elkhatib[2], V. Oyanedel-Craver[2], E. Carissimi[1]

[1]Department of Sanitary and Environmental Engineering, Federal University of Santa Maria, Av. Roraima, 1000, 97105-900, Santa Maria, RS, Brazil; gustavohbracher@gmail.com

[2]Environmental Engineering Laboratory, University of Rhode Island, Pastore 312A, Kingston, RI 02881, USA;

Keywords: Electrocoagulation, aluminum electrodes, microplastics, wastewater treatment

Abstract

The presence of microplastics (MP) in water resources has been widely reported worldwide and the negative effects to aquatic organisms, ecosystems and human health are a growing concern. Municipal wastewater is an important route for the transport of MP to the water, thus the development of techniques for MP removal in the municipal wastewater treatment is essential. In this context, the objective in this study was to evaluate an innovative semi-continuous electrocoagulation system for the MP removal from municipal wastewater. To achieve this goal, the best treatment conditions of an electrocoagulation system were evaluated using a rotatable central composite experimental design 2^2, considering the operational parameters of electrical current and electrolysis time. The electrocoagulation system was composed of a peristaltic pump, a direct-current power supply and an electrocoagulation reactor with a serpentine flow configuration, a work volume of 0.7 L and five electrodes pairs of aluminum (1.4 x 5 cm). The system was operated with a workflow of 0.9 L/min and hydraulic retention time of 22.5 min (with recirculation). The performance of the system was evaluated considering the effect on the response variables such as MP removal, electrodes mass consumption and energy consumption. The electrocoagulation system showed a removal efficiency of 96% with an electrodes mass consumption of 1.4 g/g MP and an energy consumption of 0.016 kWh/g MP. Results showed that the use of lower electrical current and higher electrolysis time (lower current density) promoted a higher energetic efficiency in the electrocoagulation system and lower electrodes mass consumption in comparison with the evaluated in the opposite conditions. So, it was possible to conclude that the electrocoagulation system presented potential to be applied for the MP removal in a municipal wastewater facility.

1 Introduction

Microplastics (MP) are defined as small plastic particles or fragments, with a diameter up to 5 mm [1]. Additionally to the exponential growth of plastic production since the 1950s, the ineffective management of end-of-life of plastic wastes turned MP presence in the environment a global problem [2]. Due to its small size, it has been proposed that MP can be easily transported in the food web of aquatic ecosystems reaching the humans by the consumption of aquatic organisms, as seafood [3]. There is a growing concern about the risks from the MP exposure to the ecosystem and the human health, because of the toxic plastics additives in the plastic manufacture such as flame retardants [3] and plasticizers [4], and the MP potential to transport chemicals as persistent organic pollutants [5, 6] and heavy metals [7].

The MP presence in water resources has been widely reported around the world [8-11], and the municipal wastewater is an important route of MP transport to aquatic environments, such as textile fibers released during washing and MP from personal care and cosmetic products [12-14]. This contamination route is even more relevant in developing countries [2], as Brazil, where only 45% of the sewage receives some kind of treatment [15] and 14% of the Brazilian cities achieve the minimum levels of treatment required by Brazilian Regulations [16]. This scenario of lack of sewage treatment associated to the issues concerning the MP presence in the water turns the study of a complete treatment system, including MP removal, extremely necessary in effluent facilities.

Electrocoagulation is an electrochemical technique that has been shown to be effective for the removal of contaminants from municipal wastewater [17, 18] and to demonstrate technical-economic advantages, as easy operation, small footprint, short operational times and low sludge production [18, 19]. Recently, the potential of the use of this technique for the MP removal from wastewater was reported [20], nevertheless the available studies were carried out only in batch configurations at lab scale, and so, there is an absence of studies exploring continuous reactor designs and configurations, which are the most viable options in large scale applications. Thus, the main goal in this study was to evaluate the use of an innovative semi-continuous electrocoagulation system for the MP removal from municipal wastewater. This study is the first, considering the techno-scientific publications, to explore a continuous configuration for an electrocoagulation system for MP removal.

2 Materials and Methods

Wastewater sample

Synthetic municipal wastewater spiked with MP was used in this study. Commercial Red Pearl Brilliant (909 RB) polyester glitter powder with particle sizes varying between 20 to 150 μm was used as model MP. The size range of the model MP represents very small particles [21], which removal by phase separation is difficult. A concentration of 25 mg/L of MP was used for all experiments. This concentration was adopted to obtain a better detection of the MP removal. The electrical conductivity of the solution was adjusted to 650 μS/cm with the addition of 300 mg/L of NaCl and the pH 7 was fixed with the addition of NaOH and HCl; these values were adopted according to common electrical conductivity and pH ranges reported for municipal wastewater [22].

Electrocoagulation system

The study was carried out in an electrocoagulation system composed of an electrocoagulation reactor, a direct-current power supply and a peristaltic pump, as showed in Figure 1. The electrocoagulation reactor was made of polyvinyl chloride and had a work volume of 0.7 L, a serpentine flow configuration and five electrodes pairs of aluminum (1.4 x 5 cm), connected in parallel and allocated at the upflow parts of the reactor, with a distance of 0.3 cm between the cathodes and anodes of each pair. The system was designed with a workflow of 0.9 L/min, hydraulic retention time of 22.5 min (with recirculation) and different electrolysis time and electrical current.

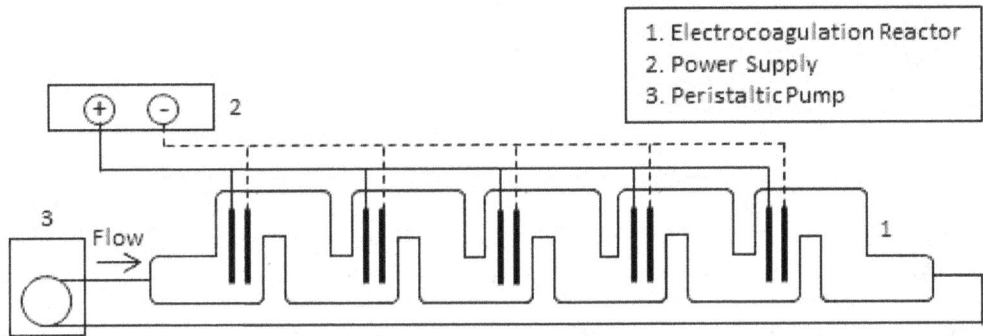

Figure 1: Electrocoagulation system

Experimental procedures

The MP removal from municipal wastewater was evaluated as function of the electrolysis time and electrical current adopted in the electrocoagulation system, by means of a rotatable central composite experimental design (RCCD) 2^2. In this experimental design, the electrolysis time and the electrical current were studied in five levels (Table 1), with three replications of the center point, which resulted in a total of 11 assays. After the detention time of 22.5 min, the samples were collected and allowed to stand for 15 min for the formed flocs flotation and sedimentation.

Table 1: Independent factors and variation levels evaluated and the assays performed in the rotatable central composite experimental design 2^2

Independent factor	Variation level				
	- 1.41	- 1.00	0.00	+ 1.00	+ 1.41
Electrical current (A)	0.02	0.08	0.24	0.39	0.45
Electrolysis time (min)	1.00	4.15	11.75	19.35	22.50

In parallel to the MP removal, the energy consumption and electrodes mass consumption were also evaluated as response variables in order to obtain the input required for the MP removal at the different electrolysis time and electrical current conditions. From the results obtained in the experimental design, regression models and response surfaces were generated with the software Statistica® 7.0 (Statsoft Inc., Tulsa, OK, USA).

Analytical procedures

The MP removal was estimated by the turbidity measurement of the samples before and after each assay, as shown in the equation 1. The turbidity samples were analyzed with a turbidimeter Orion AQ4500, according to the protocol 2130 B of the Standard Methods for the Examination of Water and Wastewater [23].

$$MP_{Removal} = \frac{T_i - T_f}{T_i} \cdot 100 \qquad \text{Eq. 1}$$

Where $MP_{Removal}$ is the MP removal (%), T_i is the sample turbidity before the treatment (NTU) and T_f is the sample turbidity after the treatment (NTU).

The energy consumption was determined according to the equation 2 by controlling electrical voltage, electrical current and electrolysis time in each assay.

$$C_{energy} = \frac{U \cdot i \cdot t}{V} \qquad \text{Eq. 2}$$

Where C_{energy} is the energy consumption (Wh/m^3), U is the electrical voltage (V), i is the electrical current (A), t is the electrolysis time (h) and V is the treated effluent volume (m^3).

The electrodes mass consumption was obtained from the weighting of the electrodes before and after each assay in an analytical balance Bel M214AIH, according to the protocol 2540 B of the Standard Methods for the Examination of Water and Wastewater for total solids determination [23].

3 Results and Discussion

The results obtained from the experimental design are shown in the Table 2. The MP removal efficiencies were greater than 72% under all conditions studied, which proves the efficacy of the electrocoagulation technique for MP removal from wastewater.

The regression models generated from the results and expressed in the Table 2 were described by the equations 3, 4 and 5, for the response variables MP removal, electrodes mass consumption and energy consumption, respectively. As the Table 3 shows, all models presented a calculated F (F_{calc}) greater than the tabulated F (F_{tab}), and a determination coefficient (R^2) greater than 0.91, which indicated that the models were statistically significant for a confidence interval of 95% and presented a good adjust to the results obtained.

$$MP_{Removal} = 94.39 + 6.16 \cdot i - 3.72 \cdot i^2 + 8.97 \cdot t - 4.48 \cdot t^2 - 2.79 \cdot i \cdot t \qquad \text{Eq. 3}$$

$$C_{Electrodes} = 25.96 + 16.70 \cdot i - 0.07 \cdot i^2 + 16.70 \cdot t - 0.07 \cdot t^2 + 10.80 \cdot i \cdot t \qquad \text{Eq. 4}$$

$$C_{Energy} = 0.34 + 0.42 \cdot i + 0.11 \cdot i^2 + 0.31 \cdot t + 0.03 \cdot t^2 + 0.36 \cdot i \cdot t \qquad \text{Eq. 5}$$

Where $MP_{Removal}$ is the MP removal (%), $C_{Electrodes}$ is the electrodes consumption (g/m³), C_{Energy} is the energy consumption (kWh/m³), i is the electric current (codded value) and t is the electrolysis time (codded value).

Table 2: Assays performed and results obtained from the rotatable central composite design 2^2

Assay	Electrical Current (A)[a]	Electrolysis time (min)[a]	Microplastics Removal (%)	Electrodes Mass Consumption (g/m³)	Energy Consumption (kWh/m³)
1	0.08 (-1)	4.15 (-1)	74.8	3.2	0.025
2	0.08 (-1)	19.35 (+1)	94.3	15.0	0.118
3	0.39 (+1)	4.15 (-1)	88.5	15.0	0.289
4	0.39 (+1)	19.35 (+1)	96.8	70.0	1.814
5	0.02 (-1.41)	11.75 (0)	72.8	2.2	0.009
6	0.45 (+1.41)	11.75 (0)	96.3	49.4	0.964
7	0.24 (0)	1.00 (-1.41)	67.5	2.2	0.030
8	0.24 (0)	22.50 (+1.41)	98.6	49.4	0.646
9	0.24 (0)	11.75 (0)	94.2	25.8	0.331
10	0.24 (0)	11.75 (0)	94.0	25.8	0.350
11	0.24 (0)	11.75 (0)	95.0	26.3	0.350

[a] Coded values of variation levels in parentheses.

Table 3: Analysis of variance for the models obtained from the rotatable composite design 2^2 (confidence interval of 95%)

Response variable	Source of variation	Sum of squares	Degrees of freedom	Mean of squares	F_{calc}	F_{tab}	R^2
$MP_{Removal}$	Regression	1,127.1	5	225.4			
	Residue	115.5	5	23.1	9.76	5.05	0.91
	Total	1,242.6	10	124.3			
$C_{Electrodes}$	Regression	4,928.6	5	985.7			
	Residue	0.13	5	0.03	37,971.23	5.05	0.99
	Total	4,928.7	10	492.9			
C_{Energy}	Regression	2.72	5	0.54			
	Residue	0.16	5	0.03	16.93	5.05	0.94
	Total	2.89	10	0.29			

The response surfaces obtained for each response variable, from the regression models generated, are shown in Figure 2. Figure 2a shows that the increase of both the electrolysis time and the electrical current resulted in an increase of the MP removal in the electrocoagulation system. This increase mainly occurs as a consequence of the higher aluminum concentration and microbubbles applied at higher electrolysis times and electrical currents [24]. In the assays with the shortest electrolysis times and electrical currents, there is a higher efficiency of MP removal per gram of electrode mass consumed (Figures 2a and 2b) compared with the assays with higher electrolysis times and electrical currents. The same trend was observed by Bukhari [25] for turbidity removal from urban wastewater, and by Perren [20] for microbeads removal from a synthetic wastewater, both researches conducted with the electrocoagulation technique. These studies proposed that the particles removal mechanisms occurred in two phases: reactive and steady. During the reactive phase, charge neutralization and adsorption are the main mechanisms of particles removal, so the increase of particles removal is proportional to the increase of coagulants concentration. Subsequently, during the steady phase, particles removal is no longer proportional to the coagulant concentration, since flocculation mechanisms such as sweep-floc process are predominant [20, 25].

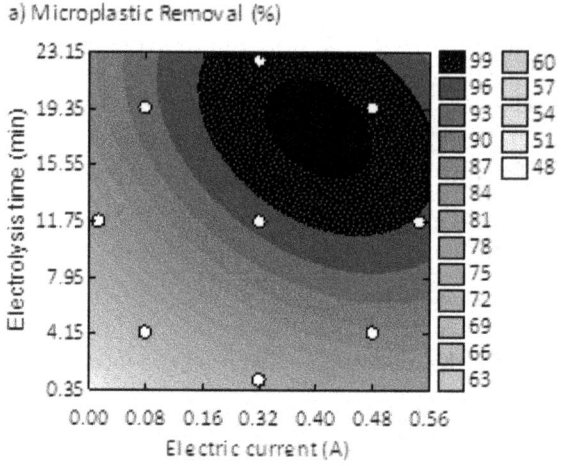

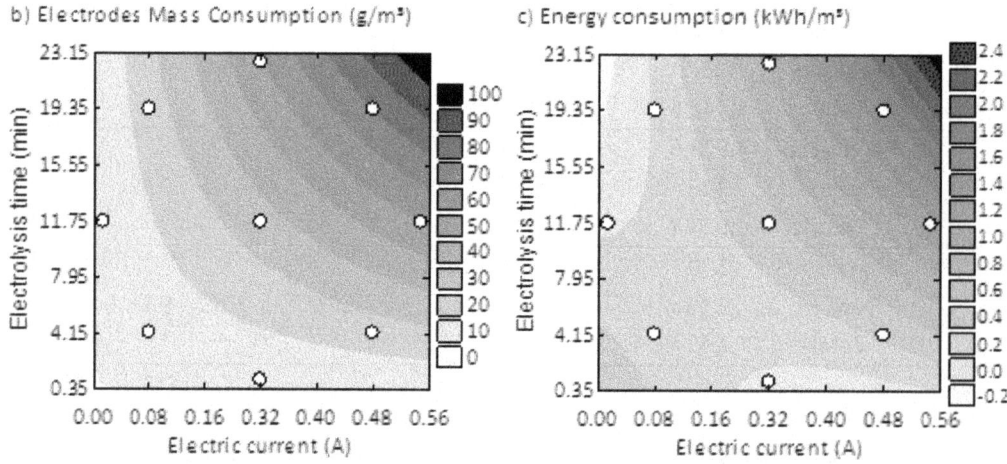

Figure 2: Response surfaces generated for the variables (a) MP removal, (b) electrodes mass consumption and (c) energy consumption, as function of the electrical current and electrolysis time

The Figures 2a and 2b show that the use of higher electrolysis times with lower electric currents is an interesting strategy to increase the MP removal efficiency. At these conditions, a greater MP removal efficiency was observed in comparison with conditions with smaller electrolysis times and greater electrical current, with the same electrodes mass consumption. This fact is related to the difference on the current density maintained during the different conditions listed above, where the current density is the quantity of electrical current established per unity of electrodes active area [26]. The assay 2 presented a removal efficiency 5.8% greater than the assay 3, and the unique difference between these assays was the current density (resulted from the electrical current and electrolysis time applied in each assay), which was 2.29 mA/cm² in the assay 2 and 11.14 mA/cm² in the assay 3. A similar difference (5.3%) was observed between the MP removal efficiency of the assays 5 and 7, where the difference was also the current density, which were 0.57 mA/cm² and 6.86 mA/cm², respectively. Comparing the assays 6 and 8, there was also a higher MP removal efficiency in the assay with lower current density (assay 8), with the same aluminum consumption, nevertheless the difference between the removal efficiencies was slightly lower in this case (2.3%). Similar results were reported by Liu [27], who observed higher removal efficiencies of diclofenac and ketoprofen from hospital wastewater by electrocoagulation with aluminum electrodes at lower current density. According to the authors, the higher removal efficiencies at lower current density occurred, because the adsorption was the main removal mechanism for the pharmaceuticals by electrocoagulation, and this mechanism could be favored with lower aluminum release rate [27]. So, the higher MP removal at lower current densities may be a result of the predominance of the adsorption on the MP removal by electrocoagulation. Consequently, the lower difference on MP removal in the assays 6 and 8 may be related to the higher aluminum concentration applied in these assays, which promotes the parallel occurrence of flocculation mechanisms.

The use of higher electrolysis times combined with lower electrical current was also efficient to obtain a more energetic efficiency removal of MP, as shown in the Figures 2a and 2c. In addition, Figures 2b and 2c show that there was a more energetic efficiency for the aluminum release in these conditions, since the same aluminum concentrations were achieved with lower energy consumption in these conditions. The energetic efficiency in these conditions was also a result of the low current density maintained in these assays. This phenomenon was also observed by Santiago [28], who reported lower electrical energy requirements for arsenic removal and aluminum release by electrocoagulation with lower current density. Similar results were obtained by Hooshmandar [29], where the electrocoagulation system presented dye removal with lower energy and aluminum consumption at the lowest current density. According to Deshpande [30], at high current density, there was a higher energy lost in water hydrolysis reactions and heating due the Joule effect, which explains the higher energetic efficiency with lower current densities.

As showed in Figure 2a, there was a reduction in the MP removal efficiency after reaching a removal of 96%, which characterizes the beginning of a steady phase in the electrocoagulation. Thus, removal values higher than 96% would not be efficient in terms of increase of energy and electrodes mass consumption (Figures 2b and 2c). Considering this fact and the benefits of the use of higher electrolysis times with lower electrical currents, the best conditions for the electrocoagulation system was established from a desired MP removal of 96% with the highest electrolysis time studied (22.50 min). Table 4 summarizes the best conditions for the electrocoagulation system studied and the predicted results from these treatment conditions (obtained from the regression models used for the response surfaces generation).

Table 4: Best treatment conditions and the predicted results for the MP removal in the electrocoagulation system studied

Treatment conditions	
Inter-electrodes distance (cm)	0.3
Flow (L/min)	0.9
Electrical current (A)	0.16
Electrolysis time (min)	22.5
Current density (mA/cm^2)	4.57
Predicted results	
MP removal (%)	96.0
Electrodes mass consumption (g/m^3)	33.1
Energy consumption (kWh/m^3)	0.39
Electrodes mass consumption (g Al/g MP)[a]	1.4
Energy consumption (kWh/g MP)[b]	0.016

[a] Gram of aluminum electrodes consumed per gram of MP removed.
[b] kWh consumed per gram of MP removed.

The energy consumption value of 0.016 kWh/g MP was slightly higher than the batch electrocoagulation system used by Perren [20], who reported minimum energetic consumptions of 0.05 g MP/kJ (0.0056 kWh/g MP). However, Perren [20] worked with a conductivity 11 times higher than those used in this study, implying in high treatment costs due the need for the addition of electrolytes and need for conductivity adjustment for the wastewater treatment. Therefore, the best conditions found in this study are more energetically efficient than those used by Perren [20].

The electrocoagulation system studied also presented a better efficiency in comparison with the conventional chemical coagulation with aluminum salts addition. Ma [31] reported aluminum mass consumptions higher than 2.6 g Al/g MP in a batch water treatment by coagulation, flocculation and sedimentation without auxiliary flocculant use, and aluminum mass consumption higher than 1.1 g Al/g MP with the addition of 15 g/m^3 of anionic polyacrylamide.

Considering the market costs of the aluminum electrodes (R$ 0.022/g Al) and of the electricity (R$ 0.5425/kWh) in Brazil, the electrocoagulation system presented an operational cost of R$ 0.04 (equivalent to US$ 0.01) per gram of MP removed. The main part of the costs was a result from the electrode mass consumption, which represents 78% of the operational costs.

4 Conclusions

This study presented a novel design for electrocoagulation treatment systems in continuous flow. The electrocoagulation system proposed presented potential to be applied for the MP removal from municipal wastewater. The electrocoagulation system presented a MP removal efficiency of 96% with an electrodes mass consumption of 1.4 g/g MP and an energy consumption of 0.016 kWh/g MP. The adoption of lower current density in the electrocoagulation proved to be a good alternative to obtain higher energetic efficiency in the electrocoagulation system and lower electrodes mass consumption.

5 Acknowledgements

Mr. Gustavo Holz Bracher would like to acknowledge the German Academic Exchange Service (DAAD) and Exceed Swindon project for support his participation at the event "Microplastics in the Water Environment" held in Koh Samui, Thailand on August 19-21, 2019.

6 References

[1] GESAMP: Sources, fate and effects of microplastics in the marine environment: a global assessment, IMO/FAO/UNESCO-IOC/UNIDO/WMO/IAEA/UN/UNEP/UNDP Joint Group of Experts on The Scientific Aspects of Marine Environmental Protection, London (2015).

[2] UNEP: Marine plastic debris and microplastics – Global lessons and research to inspire action and guide policy change, United Nations Environment Programme, Nairobi (2016).

[3] M. Smith, D.C. Love, C.M. Rochman, R.A. Neff: Microplastics in seafood and the implications for human health, Current Environmental Health Reports 2018, 5, 375-386.

[4] J. Bayo, S. Olmos, J. López-Castellanos: Non-polymeric chemicals or additives associated with microplastic particulate fraction in a treated urban effluent, WIT Transactions on The Built Environment 2018, 179, 303-314.

[5] L.M. Rios, P.R. Joes, C. Moore, U.V. Narayan: Quantification of persistent organic pollutants adsorbed on plastic debris from the Northern Pacific Gyre's "eastern garbage patch", Journal of Environmental Monitoring 2010, 12, 2226-2236.

[6] E. Chua, J. Shimeta, D. Nugegoda, P.D. Morrison, B. Clarke: Assimilation of polybrominated diphenyl ethers from microplastics by the marine amphipod, *Allorchestes compressa*, Environmental Science & Technology 2014, 48(14), 8127-8134.

[7] C.M. Rochman, B.T. Hentschel, S.J. Teh: Long-term sorption of metals is similar among plastic types: implications for plastic debris in aquatic environments, PLoS One 2014, 9(1), e85433.

[8] A.D. Gray, H. Wertz, R.R. Leads, J.E. Weinstein: Microplastic in two South Carolina Estuaries: Occurrence, distribution, and composition, Marine Pollution Bulletin 2018, 128, 223-233.

[9] S. Klein, E. Worch, T.P. Knepper: Occurrence and spatial distribution of microplastics in river shore sediments of the Rhine-Main area in Germany, Environmental Science & Technology 2015, 49, 6070-6076.

[10] R.S. Pazos, T. Maiztegui, D.C. Colautti, A.H. Paracampo, N. Gómez: Microplastics in gut contents of coastal freshwater fish from Río de la Plata estuary, Marine Pollution Bulletin 2017, 122, 85-90.

[11] K. Zhang, X. Xiong, H. Hu, C. Wu, Y. Bi, Y. Wu, Y. Bi, Y. Wu, B. Zhou, P.K.S. Lam, J. Liu: Occurrence and characteristics of microplastic pollution in Xiangxi Bay of Three Gorges Reservoir, China, Environmental Science & Technology 2017, 51, 3794-3801.

[12] M.A. Browne, P. Crump, S.J. Niven, E. Teuten, A. Tonkin, T. Galloway, R. Thompson: Accumulation of microplastic on shorelines woldwide: sources and sinks, Environmental Science & Technology 2011, 45, 9175-9179.

[13] S.A. Mason, D. Garneau, R. Sutton, Y. Chu, K. Ehmann, J. Barnes, P. Fink, D. Papazissimos, D.L. Rogers: Microplastic pollution is widely detected in US municipal wastewater treatment plant effluent, Environmental Pollution 2016, 218, 1045-1054.

[14] S. Karbalaei, P. Hanachi, T.R. Walker, M. Cole: Occurrence, sources, human health impacts and mitigation of microplastic pollution, Environmental Science and Pollution Research 2018, 25, 36046-36063.

[15] SNIS: Diagnóstico dos Serviços de Água e Esgotos – 2016, Sistema Nacional de Informações sobre Saneamento / Ministério das Cidades, Brasília (2018).

[16] ANA: Conjuntura dos recursos hídricos no Brasil 2017: relatório pleno, Agência Nacional de Águas, Brasília (2017).

[17] E.M. Symonds, M.M. Cook, S.M. McQuaig, R.M. Ulrich, R.O. Schenck, J.O. Lukasik, E.S. Van Vleet, M. Breitbart: Reduction of nutrients, microbes, and personal care products in domestic wastewater by a benchtop electrocoagulation unit, Scientific Reports 2015, 5, e9380.

[18] M. Elazzouzi, K. Haboubi, M.S. Elyoubi: Electrocoagulation flocculation as a low-cost process for pollutants removal from urban wastewater, Chemical Engineering Research and Design 2017, 117, 614-626.

[19] M.M. Emamjomeh, M. Sivakumar, A.S. Varyani: Analysis and the understanding of fluoride removal mechanisms by an electrocoagulation/flotation (ECF) process, Desalination 2011, 275(1-3), 102-106.

[20] W. Perren, A. Wojtasik, Q. Cai: Removal of microbeads from wastewater using electrocoagulation, ACS Omega 2018, 3(3), 3357-3364.

[21] H.K. Imhof, N.P. Ivleva, J. Schmid, R. Niessner, C. Laforsch: Contamination of beach sediments of a subalpine lake with microplastic particles, Current Biology 2013, 23(19), 867-868.

[22] S.T. Decezaro: Septic tank and vertical flow constructed wetland system with recirculation for organic matter and nitrogen removal from domestic wastewater [dissertation], Federal University of Santa Maria, Santa Maria (2018).

[23] APHA: Standard methods for examination of water and wastewater, American Public Health Association, Washington (2012).

[24] G.H. Bracher, A.G. Glusczak, E.A. Somavilla, C. Graepin, E. Carissimi: Effect of different operational parameters in the turbidity removal from domestic wastewater by electrocoagulation-flotation, 11º Simpósio Internacional de Qualidade Ambiental, 02-04 October 2018, Porto Alegre, Brazil.

[25] A.A. Bukhari: Investigation of the electro-coagulation treatment process for the removal of total suspended solids and turbidity from municipal wastewater, Bioresource Technology 2008, 99(5), 914-921.

[26] J.N. Hakizimana, B. Gourich, M. Chafi, Y. Stiriba, C. Vial, P. Drogui, J. Naja: Electrocoagulation process in water treatment: A review of electrocoagulation modeling approaches, Desalination 2017, 404, 1-21.

[27] Y.J. Liu, S.L. Lo, Y.H. Liou, C.Y. Hu: Removal of non-steroidal anti-inflammatory drugs (NSAIDs) by electrocoagulation–flotation with a cationic surfactant, Separation and Purification Technology 2015, 152, 148-154.

[28] S.L.G. Santiago, S.P. Castrejón, A.M. Domínguez, A.M. Brito, I.E.V. Mendoza: Current density as a master variable in designing reactors, Procedia Chemistry 2014, 12, 66-72.

[29] A. Hooshmandfar, B. Ayati, A.K. Darban: Optimization of material and energy consumption for removal of Acid Red 14 by simultaneous electrocoagulation and electroflotation, Water Science and Technology 2016, 73(1), 192-202.

[30] A.M. Deshpande, S. Satyanarayan, S. Ramakant: Treatment of high-strength pharmaceutical wastewater by electrocoagulation combined with anaerobic process, Water Science and Technology 2010, 61(2), 463-472.

[31] B. Ma, W. Xue, C. Hu, H. Liu, J. Qu, L. Li: Characteristics of microplastic removal via coagulation and ultrafiltration during drinking water treatment. Chemical Engineering Journal 2019, 359, 159-167.

POTENTIAL OF PLASTIC-DEGRADING MICROBES IN PREVENTING PLASTIC POLLUTION AND ITS CIRCULAR SUSTAINABILITY PROGRAM

Desak Ketut Tristiana Sukmadewi[1], Ade Brian Mustafa[1,2]

[1]*IPB University (Bogor Agricultural University), Faculty of Agriculture, Departement of Soil Science and Land Resource, Bogor, West Java, Indonesia 16680;* tristianasukmadewi@yahoo.com

[2]*Shanghai Jiao Tong University, School of Environmental Science and Engineering, Shanghai, China 200240;*

Keywords: Plastic degrading microbes, pollution prevention potential, circular sustainability program, polyethylene

Abstract

Biotechnology offers a potentially powerful approach to the management of plastic wastes. Plastic waste could be degraded through biodegradation processes. An environmental factor that influences the biodegradation process is a microbe. This paper aims to study potential plastic degrading microbes in preventing plastic pollution and its circular sustainability program. This paper is based on literature review studies related to plastic degradation based on weight loss measurements (determination of residual polymer). To confirm the degradation process, some literature conducted further tests using Scanning Electron Microscopy (SEM) and Fourier Transform Infrared Spectroscopy (FTIR). Biodegradation process might be accelerated and efficient with the application of microbes of degradation activity. According to given data, microbes can degrade plastic, especially polyethylene (PE) with the ability to range from 8-46.7% in a period that varies from one month to six months in different media. Microbes that can degrade plastic waste were *Bacillus cereus, Pseudomonas* sp., *Aspergillus oryzae, Arthrobacter, Streptomyces sp, Aspergillus flavus, Rhodococcus ruber,* and *Penicillium simplicissimum, Brevibacillus borstelensis*. Solid polymer degradation by microbes, e.g., of PE, requires the formation of a biofilm on the surface of the polymer to enable the microorganism to efficiently utilize the non-soluble substrate. These microbes are using oxidation and hydrolysis processes by produced enzymes that leads to segregation of high polymer chain to be a small monomer by some metabolism process. The implementation of this concept can be applied by creating a bioreactor in each landfill. The circularity aspect would be stemmed from bioreactor processes, in which the microbes degrading plastics and their enzymatic products are attainable. Afterward, this material could be used as a substrate resources for microbial growths. The integration of mechanical, chemical, thermochemical and biotechnological recycling techniques with microbial ones may perhaps be the key to attaining the goal of a circular program in this sector.

1 Introduction

Indonesia is facing a waste-related state of emergency. According to data from the Ministry of Environment and Forestry of Republic of Indonesia (MoEF) [1], an average person in Indonesia produces 0.7 kg of waste per day in 2015. As many as 175,000 t of waste was produced daily, reaching 64 million t/yr. The majority of Indonesia's waste is still ended up on landfill (69%) without proper management, 15% are burnt and buried, while only 7.5% is recycled and composted. The core problem is unintegrated waste management, from upstream to downstream. A lot of waste has not been sorted at the source (upstream), and this affects the recycling industry, which has difficulty to recover waste materials (downstream). According to MoEF, 81% of the waste was not separated at the source in 2015, making it loses the value, ending up at the landfill and leaking to oceans [1].

Environment in Indonesia is in the state of emergency. It needs to be managed appropriately. According to Wendling et al. (2018), Indonesia is ranked by the score of 46.92 in the Environmental Performance Index (EPI), making this nation in the list number 133 out of 180 countries. Furthermore, Indonesia as the second biggest country in the world after China, doing its plastic waste polluting the sea [2]. The waste may accelerate worst benevolance on environment and health, therefore, some strategies and management approaches need to be taken into account. The government has also made legal commitments to provide environmental education, such as Law No. 32/2009 on Environmental Protection and Management. But, Indonesia did not do well in its 'report card' for the seventh Millennium Development Goal, promoting environmental sustainability. Indonesia needs to move forward fast in environmental education. That is why this has become an on-going home work to be able to encourage the spirit and hard work to raise the environment issue and social awareness in Indonesia.

The Indonesian government has stated a commitment to cut the volume of plastic waste in the oceans by 70% by 2025, preparing a roadmap and funding of USD 1 billion to achieve it. Circular economy is a way to solve waste problems around us and to save the resources in the future. Furthermore, Indonesia is still in its initial process to realize it. The spirit of circular economy has been manifested in the grand design of waste reduction and waste handling policy in Perpres (Presiden Regulation) No 97/2017 regarding National Strategy and Policy (Jakstranas). Jakstranas of household and similar-to-household waste treatment at national, provincial and regency level have set the target of waste reduction at 30% and waste handling at 70% by 2025 [1].

Plastic waste disposal in landfills is a common practice followed in many countries. Both "dry-tomb" and "wet-cell" landfill approaches are used for plastic waste disposal, though the latter design is more common. Pressures on landfills are building across municipalities because of a paucity of land. In addition, the wet-cell landfill approach leads to the generation of a landfill leachate, which has the potential to contaminate the environment.

Reducing plastic waste on land will have a wide impact on the volume of waste in the sea. One of them will reduce secondary microplastics (large plastic degraded into small pieces). Therefore, some strategies and management approaches need to be taken into account. Biotechnology offers a potentially powerful approach to the management of plastic wastes. The rapid growth of the world population combined with a focus on consumption rather than recycling has led to the generation of huge quantities of plastic waste and the consequent detrimental impact on landfills, waterways, and the marine environment.

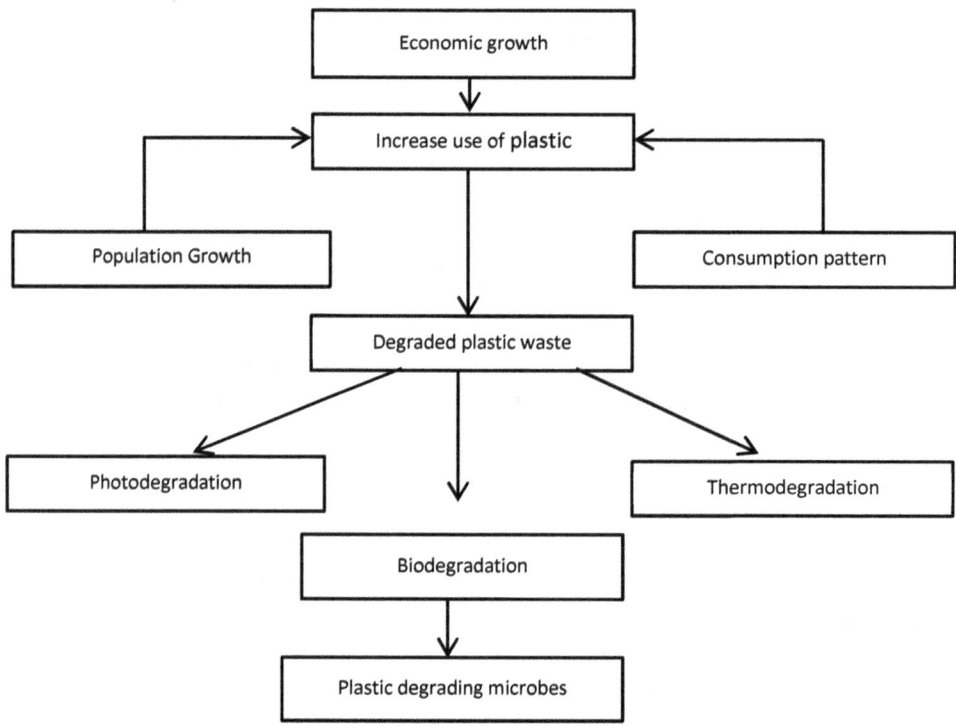

Figure 1: Flow-chart of the state of the art

This paper aims to study potential plastic-degrading microbes in preventing plastic pollution and its circular sustainability program.

2 Methods

This paper framework is based on literature studies. In this paper, a reference using weight loss measurement for determination of residual polymer for plastic degradation is utilized. The percentage of degradation by microbes were determined by calculation of the percentage of weight loss of plastics.

To confirm the occurrence of the degradation process, several authors conducted further experiments using Scanning Electron Microscopy (SEM) and Fourier Transform Infrared Spectroscopy (FTIR). These methods were mainly conducted in order to identify an alteration of the plastic structure after incubation with microbes. Several experimental methods to affirm the degradation processes were conducted as screening the enzymes responsible for plastic degradation, mass production by sub-merged fermentation, enzyme assay, protein estimation, determination of molecular weight of plastic, evaluation of bacterial hydrophobicity, estimation of microbe biomass colonizing the plastic, and estimation of microbe biofilm viability, etc.

3 Results and Discussion

3.1 Potential of plastic-degrading microbes in preventing plastic pollution

The durability of plastic is the main obstacle to its degradation in the environment. However, some degradation does occur. Biodegradation is defined as a natural process of degrading materials through microbes such as bacteria, fungi and algae [4]. Biodegradation process might be accellerated and efficient with the application of microbes with degradable-activity or in the utilization of plastic as a growth substrat. Indigeneous microbes provide several advantages in regard with specific-priority utilization in degrading plastic, ultimately for specific-based microbes from their original location. This, of course, will make the processing of plastic waste faster and more efficient.

This paper aims to contend for framing biotecnological approaches with the circular economy. Both of the implementation will lead to sustainable development, in line with maximizing knowledge for mother-nature. Microbial degradation by microbes occurres naturally in the ecosystem, however, the skyrocketed production of plastic waste tremendously need a solution that the microbiology will pave the way in this scheme (Figure 1). Previous research has revealed indigenous microbes isolated from garbage and plastic dumping site shown in Table 1. According to given data, the ability of microbes to degrade plastic could reach 46.7% by *Streptomyces sp.* *Streptomyces sp.* That is isolated from garbage soil can degraded low density polyethylene (LDPE) in a six months period in mineral salt medium (PE bag was cut into equal pieces of 5 cm x 2 cm). This capability has a prominent evidence for further exploration and research. Each microbe has a different ability to degrade plastic. Degradation efficacy varies with different times.

Table 1: Soil microbe efficacy in degrading plastics

Microbe	Type of plastic	Source of microbes	Degradation efficacy	Time (months)	Medium	Ref.
Bacillus cereus	PE	Dumping site	14%	3	Culture broth medium (50 mL), PE bags and disposable plastic cups (1 cm dia), (PE treated in UV)	[5]
Streptomyces sp.	LDPE	Garbage soil	46.7%	6	Mineral salt medium (100 mL) contains LDPE (5 cm x 2 cm)	[6]
Streptomyces KU8, *Pseudomonas* sp. and *Aspergillus flavus*	LDPE powder	Garbage soil and dumping site	46.16%, 37.09%, 20.63%	6	Culture broth medium (50 mL), PE bags and disposable plastic cups (1 cm dia)	[7]

Pseudomonas sp.	PE and plastic	Mangrove soil	20.54%, 8.16%	1	Mineral salt medium (100 mL), PE bags and plastic cups (2 cm dia)	[8]
Aspergillus glaucus	PE and plastic	Mangrove soil	20.80%, 7.26%	1	Mineral salt medium (100 mL), LDPE (5 x 2 cm)	[8]
Micrococcus luteus	Plastic cup	Forest soil	38%	2 (55 days)	Nutrient broth medium (250 mL), 1 g plastics cups cut into small pieces	[9]
Masoniella sp.	Plastic cup	Forest soil	27.4%	2 (55 days)	Nutrient broth medium (250 mL), 1 g plastics cups	[9]
Rhodococcus ruber	PE	Soil that contain PE waste	8%	1	Synthetic medium (100 mL), PE films cut into pieces of 3×3 cm (PE treated in UV)	[10]
Penicillium simplicis-simum	PE	Soil from local dumping site of Shivamogga district	38%	3	Salt medium (50 mL), PE (1 cm dia)	[11]
Brevibacillus borstelesis	PE	PE waste disposal site	11%	1	Mineral medium (100 ml), PE (3 x 3 cm each), PE treated with UV	[12]

PE: Polyethylene; LDPE: Low density polyethylene

Based on FTIR analysis [10] with treatment *Rhodococcus ruber* (C208) in PE degradation showed that the amount of carbonyl moiety in PE decreased after incubation due to their utilization by the bacterium. This is in accordance with the biodegradation mechanism suggested by Albertsson et al. (1987) [13] of a synergistic effect between photooxidation and the biodegradation of PE. The decrease in the amount of double bonds obtained after incubation with *Rhodococcus ruber* (C208) may be explained by the degradation of short PE oligomers produced during photooxidation. Microbial degradation of solid polymers like PE requires the formation of a biofilm on polymer surface in order to enable the microorganism to efficiently utilize the non-soluble substrate. Indeed, *Rhodococcus ruber* (C208) effectively colonized the PE surface. This may explain the relatively rapid biodegradation of PE, which was evident (as measured by weight loss and FTIR) as early as 2 weeks after inoculation.

Research conducted by Gilan et al. (2004) [10] provides an indication of the presence of enzymatic activity that might affect PE. SEM photomicrographs of the bacterial biofilm showed some localized degradation of the PE around the bacterial cells in the biofilm, forming a cell-like molded pattern in the polymer. Such patterns have previously been observed for biodegradable polymers, e.g., poly-β-hydroxybutyrate [14]. Protein assays and FDA hydrolysis by extracellular esterases proved to be efficient tools for determining the state of PE colonization and biofilm formation. Both methods provided strong evidence of rapid colonization of PE during the first 2 days of incubation, followed by a sharp decrease in biomass density.

According to the research of Hadad et al. 2005 [12], *B. borstelensis* was found capable of utilizing standard and photo-oxidized PE as the sole carbon source. During the 1-month incubation with *B. borstelensis*, the maximum biodegradation of PE, measured in terms of gravimetric and molecular weight losses, was *ca.* 11% and 30%, respectively. This biodegradation level is higher than the values reported for PE incubated in soil for 10 years, ranging from 3.5% to 8.4% [15].

Based on FTIR analysis, it is shown that the FTIR spectrum of UV photo-oxidized PE (LDPE-L0235) showed a typical carbonyl peak at 1712 cm^{-1}. Incubation of the photo-oxidized PE with *B. borstelensis* for 30 d showed a marked reduction in the amount of carbonyl moieties. The reduction in carbonyl moieties was also estimated in terms of carbonyl index, which is the ratio between the absorbance peak of carbonyl to that of CH_2 at 1462–1463 cm^{-1}. It was found that the incubation of the UV irradiated LDPE-L0235 with *B. borstelensis* strain 707 reduced the carbonyl index by ca 70%.

Moreover, the research of Somya et al. (2014) [11] showed that after the treatment of PE with UV light, *Bacillus cereus* was able to degrade it more efficiently. Carbonyl groups are produced by UV light or other oxidizing agents, and these functional groups are the main factors at the beginning of the degradation, being attacked by microorganisms that degrade the shorter segments of PE chains. FTIR results showed the formation of aldehyde, alcohol, carboxylic acid, aromatic, alkene and ether groups indicating the degradation of PE by isolated bacteria. Morphological alteration could be seen when observed through SEM. Formation of holes and disruption of the PE structure confirmed the degradation capacity of *Bacillus cereus*. Results confirmed the role manganese peroxidase as key enzyme in biodegradation of PE [16]. Role of laccase-mediator system was investigated for biodegradation of PE in the presence of 1-hydroxybenzotriazole (HBT). Laccase of *Trametes versicolor* was used. Degradation of PE was confirmed by changes in relative elongation, relative tensile strength and molecular weight distribution. All these results confirmed the degradation of PE by the laccase mediator system [17]. *Bacillus cereus* showed positive results for laccase indicating its possible role in PE degradation

The implementation of this concept can be realized by installing a bioreactor or biofermentor in each landfill. Bioreactors or fermentors can be used for microbial production on a large scale. The principle used in the fermentor is to ensure the growth of microbes and products from microbes in the fermentor. All parts of the fermentor must be in the same condition and all nutrients including oxygen must be available entirely. In this case, plastic serves as carbon source for the growth of microbes. The fermentor scale can be used on a large scale fermentor (such as in petrochemical industry) and have to consider for the distribution of the culture medium in the fermentor.

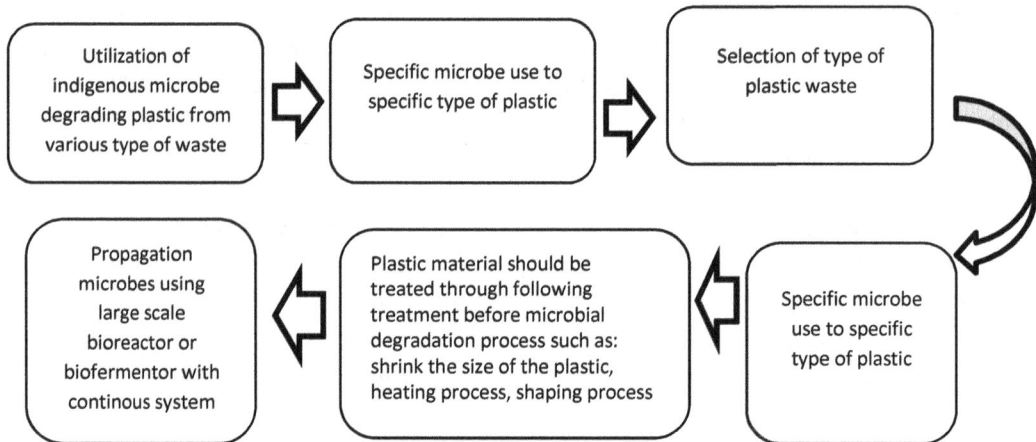

Figure 2: Implementation concept of plastic-degrading microbes

The fermentation process can be used in continues process. In this process, substrate drainage and product retrieval are carried out continuously every time after the maximum product concentration or limiting substrate is obtained to reach a concentration that is almost constant. In this case, the substrate and inoculum can be added together in a continuous manner so that the exponential phase can be extended, then more bacterial cell biomass is obtained. These bacterial culture is then used to degrade the plastic. Previously, the garbage in the landfill had been selected before being applied to microbes, which were able to degrade plastic. Biofermentor methods are related with inoculum-based exposure. Therefore, in a pure culture method, specific bacteria and fungi can be applied for degradation of polymers [16].

Various microbes such as bacteria and fungi could degrade the plastic. These microorganisms are using oxidation and hydrolysis process by producing enzymes that lead to segregation of high polymer chains to small oligomers and monomers by some metabolism process. Microbial growth is influenced by water availability, potential redox, temperature, carbon and energy sources. Microbes could generate exo-enzymes and endo-enzymes that attached in high molecules of substrate surface and become a separated small molecule (in size). Microbes are recognizing polymer as an organic resource [19]. The attachment of microbes on the PE surface means, the degradation process is beginning. Some microbes such as *Streptomyces viridosporus* T7A, *Streptomyces badius* 252, and *Streptomyces setonii* 75Vi2 and wood degrading fungi produce some extracellular enzymes, which lead to the degradation of PE. The extracellular enzymatic complex (ligninolytic system) in wood degrading fungi contains peroxidases, laccases and oxidases that lead to the production of extracellular hydrogen peroxide. Depending upon the type of the organism or strain and culture condition, the characteristics of this enzyme system were various [20, 21].

3.2 Circular sustainability programm of plastic-degrading microbe

The circularity aspect would be stemmed from bioreactor processes, in which the microbes degrading plastics and their enzymatic products are attainable. Afterward, this material could be used as a substrate resources for the microbial growths (Figure 3).

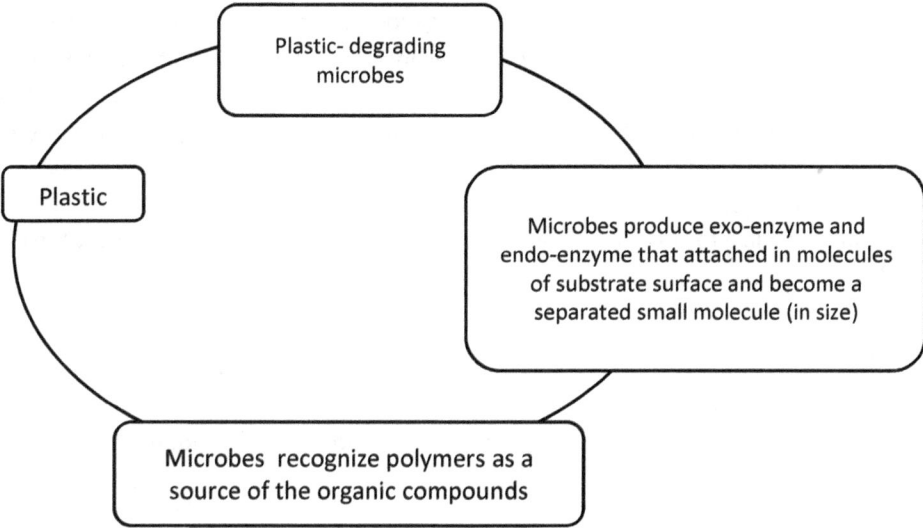

Figure 3: The interrelationship of microorganisms and sustainability issues: a circular Framework

Sustainability issues, the role of microorganisms, plastic and marine debris are in interconnectedness. The circular economy focuses on the retention of value in the material cycle through the re-use and recycling of materials such as plastic and preventing their uncontrolled release into the environment. The circular economy focuses on the recycling of plastics. A current gap in circular economy thinking with respect to plastics is the role of biodegradable plastics in the circular economy. Biodegradable plastics such as polylactic acid can be mechanically recycled but also converted by biological processes, where carbon can be returned to nature in a safe and sustainable way, e.g., by composting, which is central to a circular economy. The combination of pyrolysis and microbiology to convert non-degradable plastics into biodegradable ones was also reported, offering an unconventional route for non-degradable plastics to cross over from the technical half of the circular economy to the biological half. If improvements to enzyme activity are made, then one can envisage enzyme technologies entering the technical half (i.e., pyrolysis and depolymerization of plastics) of the circular economy in the future. The generated monomers could be used to make more biodegradable plastics, which would represent a complete biological recycling of plastics in a circular economy. The integration of hydrolytic enzymes into a microbial chassis would result in a customized microbial platform that is capable of converting plastic into biodegradable counterparts in a single cell.

4 Conclusions

In conclusion, microbes have the potential to degrade plastics even if they are not up to 50%. Microbes have further opportunities to be developed by utilizing genetic engineering that can modify the ability of microbes. This microbial treatment can be a treatment, before plastic waste is disposed of freely. It possible to design efficient microbial communities able to degrade plastic waste, even those types currently recalcitrant to biologically driven breakdown. The integration of mechanical, chemical, thermochemical and biotechnological recycling techniques with microbial, fungal and even protist biological activity allowed to proceed under controlled and contained conditions, may perhaps be the key to attaining the goal of a circular program in this sector.

5 Acknowledgements

The authors would like to thank DAAD and the Exceed Swindon project to support their participation at this workshop.

6 References

[1] Ministry of Environment and Forestry Republic Indonesia: Redefining waste management : from waste to resources management. Event report the 2ND Indonesia circular economy forum 2018.

[2] Jambeck, J.R., Geyer, R., Wilcox, C., Siegler, T. R., Perryman, M., Andrady, A., Narayan, R., Law, K.L.: Plastic waste inputs from land into the ocean. Science 2015, 347, 768-771.

[3] The Central Bureau of Statistics Indonesia: Waste Management in Indonesia. Indonesian Environmental Statistics 2018.

[4] Rutkowska, M., Heimowska, A., Krasowska, K., Janik, H.: Biodegradability of Polyethylene Starch Blends in Sea Water. Pol J Environ Stud 2002, 11, 267-274.

[5] Sowmya, H.V., Ramalingappa, Krishnappa, M., Thippeswamy: Biodegradation of Polyethylene by *Bacillus cereus*. Advances Polym. Sci. Technol. An. Int. J., 2014, 4(2), 28-32.

[6] Deepika, S., Jaya, M.R.: Biodegradation of low density polyethylene by micro-organisms from garbage soil. J. Exp. Bio. Agri. Sci. 2015, 3(1), 15-21.

[7] Usha, R., Sangeetha, T., Palaniswamy, M.: Screening of Polyethylene Degrading Microorganisms from Garbage Soil. Libyan Agric. Res. Cen. J. Int. 2011, 2,200-204.

[8] Kathiresan, K.: Polythene and plastic degrading microbes from mangrove soil. Rev. Biol. Trop. 2003, 51, 629-640.

[9] Sivasankari, S., Vinotha, T.: In Vitro Degradation of Plastics (Plastic Cup) Using *Micrococcus Luteus* and *Masoniella* Sp. Sch. Acad. J. Biosci. 2014, 2(2), 85-89.

[10] Gilan, I., Hadar, Y., Sivan, A.: Colonization, biofilm formation and biodegradation of polyethylene by a strain of *Rhodococcus ruber*. Appl. Microbiol. Biotechnol. 2004, 65, 97–104

[11] Sowmya, H.V., Ramalingappa, Krishnappa, M., Thippeswamy, B.: Degradation of polythylene by *Penicillium simplicissimum* isolate from local dumpsite of Shivamogga district. Environ. Dev. Sustain. 2014.

[12] Hadad, D., Geresh, S., Sivan A.: Biodegradation of polythylene by the thermophilic bacterium Brevibacillus borstelensis. J. Appl. Microbiol. 2005, 98, 1093-1100.

[13] Albertsson, A.C., Andersson, S.O., Karlsson, S.: The mechanism of biodegradation of polyethylene. Polym. Degrad. Stabil. 1987, 18, 73–87.

[14] Otake, Y., Kobayashi, T., Ashabe, H., Murakami, N., Ono, K.: Biodegradation of low-density polyethylene, polystyrene, polyvinyl- chloride, and urea-formaldehyde resin buried under soil for over 32 years. J. Appl. Polym. Sci. 1995, 56, 1789–1796.

[15] Albertsson, A.C., Karlsson, S.: The influence of biotic and abiotic environments on the degradation of polyethylene. Prog. Polym. Sci. 1990, 15, 177–192.

[16] Iiyoshi, Y. Tsutsumi, Y., Nishida, T.: Polyethylene degradation by lignin degrading fungi and manganese peroxidase. J. Wood Sci. 1998, 222-229.

[17] Fujisawa, H., Hirai, H., Nishida, T.: Degradation of Polyethylene and Nylon-66 by laccase mediator system. J. Polym. Environ. 2001, 9, 103-108.

[18] Singh, B., Sharma, N.: Mechanistic implication of plastic degradation. Polym. Degrad. Stabil. 2008, 561-584.

[19] Raziyafathima, M., Praseetha, P.K,, Rimal, I.R.S.: Microbial Degradation of Plastic Waste: A Review. J. Pharm. Chem. Biol. Sci. 2016, 4(2), 231-242.

[20] Shah, A.A., Hasan, F., Hameed, A., Ahmed, S.: Biological degradation of plastics: A comprehensive review. Biotech. Adv. 2008, 26, 246-265.

[21] Sangale, M.K., Shahnawaz, M., Ade, A.B.: A Review on Biodegradation of Polyethylene: The Microbial Approach. J. Bioremed. Biodeg. 2012, 3, 1-9.

MICROPLASTICS CHALLENGES AND INTERNATIONAL GOVERNANCE MECHANISMS

Muhammad Mumtaz[1], Jose A. Puppim de Oliveira[2], Ishfaq Ahmed[3]

[1,2] Sao Paulo School of Management, Getulio Vargas Foundation, Brazil;
House 393, Street 59, G-10/4, Islamabad, Pakistan; mumtaz86@hotmail.com

[3] Centre for Climate Research and Development, COMSATS University Islamabad, Pakistan

Keywords: Adaptive governance, local governments, microplastics, transboundary issues

Abstract

This study contributes to understand plastic governance and analyzes governance initiatives to address plastic pollution. Plastic pollution has become a challenge for the world especially for marine life and steps are being taken from global to the local scale to address this issue. However, there are multiple challenges and policy implementation gaps to effectively address the issue of plastic pollution. This study is conducted to understand the governance problems and to uncover the implementation challenges. The study also identifies the drivers behind the existing challenges for plastic governance. The study is conducted by employing literature review as research methodology. The study finds that transboundary and multi-level governance approaches are in place to manage the issue. The study finds that the focus of governance initiatives remained mainly at regional or national scale during the first decade of this century. However, on the heels of Paris Agreement 2015, the role of local governments and civil society are recognized to deal with the issue of climate change including plastic pollution. This shift of governance to the local scale gave birth to adaptive form of governance. This governance strategy is equipped to involve all the relevant local stakeholders including media and civil society organizations and to mobilize youth to tackle the issue of plastic pollution. Despite these positive governance initiatives, certain challenges for the existing governance mechanisms are identified. Lack of cooperation among transboundary communities, different political interests, weak institutional capacity and scarcity of financial resources especially in developing world are some of the major challenges of plastic governance. The main drivers behind these challenges are identified as: regional conflicts, absence of regulatory measures, weak institutional capabilities, lack of intergovernmental cooperation, and absence of engaging local actors in decision making. It is imperative to take variety of actions, legislative measures and cooperative approaches will ultimately help to resolve this tragedy of plastic pollution.

1 Introduction

Today, plastic pollution is one of the serious environmental challenges. World plastic production has significantly increased in the recent decade to 311 million tons in 2014 as compared with 1.7 million tons in 1950 [1]. It is projected that it will reach about 1,800 million tons in 2050 [2]. It is common to found plastics in the environment due to increase in production and durability of plastics. In initial scientific and public debates, larger plastic debris were focused and investigated [1]. However, in 1970s the small plastic particles were discussed and debated [3]. In recent years, microplastics (MPs) in the environment are recognized and received attentions in scientific and policy research. Presently, MPs issue is considered an emerging area of research.

MPs are identified as a subset of marine pollution and considered an issue of increasing concern [4]. In 2004, "Microplastics" is a collective term and was firstly proposed by Thompson et al. [5]. MPs are referred to as synthetic organic polymer particles having a size smaller than 5 mm. According to the US National Oceanic and Atmospheric Administration, MPs are defined as all plastic particles or debris, which are smaller than 5 mm in diameter [6–8]. Many studies consider MPs as plastic particles with a size of less than 5 mm length, but there is no proper consensus about the lower limit. Some researchers adopt 0.5 or 1 mm as a lower limit of MPs [9]. In relevant literature, most of published data consider plastic particles ranging 1-5 mm as MPs [10, 11].

It is projected that the MPs are increasing in comparison with the large plastic debris [12, 13]. The MPs are generally found in personal care and cosmetic products, or they appear due to environmental degradation and subsequent fragmentation of large pieces of plastics by physical, chemical and biological processes [14]. MPs can be found abundantly in every corner of our oceans, which are transported by the hydrodynamic process, winds, and ocean currents over the past few years [15, 16].

There are broadly two categories of MPs in the form of primary and secondary MPs. Primary MPs are originally and purposefully manufactured for industrial and domestic applications within microscopic size [17, 18]. Primary MPs are widely used in cosmetic formulations [19, 20], such as hair coloring, nail polish, shower gels, eye shadow, and personal care products like facial cleansers, toothpastes, and air-blasting [21–23]. Secondary MPs are formed due to the degradation of macroplastics [24]. Secondary MPs are the breakdown of macroscopic plastic debris from both particles and fibres [20]. Various studies have been conducted to identify different sources of primary and secondary MPs to the overall MPs levels in the marine environment [25–27]. It is reported that land-based sources are responsible for 75–90% for plastic in aquatic environment, while 10–25% are from ocean-based sources [14].

It is estimated that primary MPs represent a small fraction of the estimated overall environmental MPs load [9]. Therefore, it is relatively easy to address and to reduce the primary MPs. However, secondary MPs are considered as the main source of MPs pollution that are uncontrolled processes such as abrasion and degradation of larger plastic products and fragments [28]. The related literature dictates that the majority of MPs in the marine environment comes from secondary sources [29].

MPs have become a serious threat to the environment. They are negatively impacting the marine organisms. Therefore, there is a relationship of human health and food security with MPs, as people eat filter-feedings marine delicacies such as shrimp, scallops, mussels and sea cucumbers [30]. Apart from impacts on biodiversity, anthropogenic debris or litter, they also have economic implications. For instance, due to increase in coastal debris loads in South Korea after heavy rainfall, tourism revenue losses are estimated 29–37 Million USD [31]. In coastal California, visitors travel longer distances due to more waste on beaches [32]. It is reported from Brazil that 85% of beachgoers will avoid beaches with high litter loads [33]. According to the Climate Change Minister of Pakistan, recent flooding in urban city of Karachi in Pakistan is caused by the MPs.

To address the challenge of MPs, the use and subsequent release of MPs must be drastically reduced by incorporating global to local initiatives. Multiple governance measures are taken to deal with this challenge. This study is conducted to review and to analyze the governance initiatives towards tacking plastic pollution. The study also identifies the drivers behind the existing challenges for plastic governance.

2 Evolution of Plastic Governance

Like other environmental challenges, plastic pollution is a complex and transboundary phenomenon and, therefore, the governance mechanisms are also complex. It is a global issue, and when it lies in areas beyond national jurisdiction, then the issues of responsibility become challenging [30]. This complex situation needs to be addressed at global level as well as at the local scale.

Solutions to successfully address the tragedy of plastic pollution requires a combination of multiple initiatives in the form of regulation, economic/market and community-based measures [34, 35]. These solutions can be from local community actions to global efforts [30]. Although, no particular international marine legislation exists with respect to tackling the problems of MPs so far, there are many proactive actions are being taken in the form of legally bindings and voluntary approaches at international, regional, national, and local levels.

There are majorly three global international conventions to manage the issue of plastic waste in the marine environment in the early 1970s: United Nations (UN) Convention on the Law of the Sea (UNCLOS, 1982), the International Convention for the Prevention of Pollution from Ships (1973) as modified by the Protocol of 1978 (MARPOL 73/78) and the Convention on the Prevention of Pollution by Dumping of Wastes and Other Matter (London Convention or LC, 1972) [36, 37].

Manila Declaration in 2012 was adopted by the European Commission and other 64 countries to prevent marine litter from land-based and sea-based sources [38, 39]. The issue of plastic waste and MPs was included into the Basel Convention at the 13th Conference of Parties in 2017 in order to encourage and to engage regional and sub-regional centers to curb the impact of plastic waste, marine plastic litter, and MPs. Some other important initiatives are also taken in recent years, such as the World Oceans Summit (2017) and at recent meetings of the top seven and top 20 global economies G7 and G20. Furthermore, a Ministerial Declaration "Toward a Pollution Free Planet" was adopted by consensus by the UN Environment Assembly (2017). Table 1 shows the legislative measures, regulations, and instruments related to MPs at international level.

Table 1: Legislation, regulations, and instruments related to MPs at international level [16]

International instruments	Time	Specific contents
United Nations Convention on the Law of the Sea MARPOL 73/78	1972	Part XII (Articles 192–237): protection and control of marine pollution from sea-/ land-based sources.
	1973	Annex V prohibits "the disposal into the sea of all plastics, cargo residues, fishing gear including but not limited to synthetic ropes, synthetic fishing nets and plastic garbage bags". (revised in 2011 and come into force in 2013)
London Convention	1972	To prevent the "deliberate disposal at sea of wastes and other matter from vessels, aircraft and other structures, including the vessels themselves". (Annex I, paragraph 2)
London Protocol	1996	To prohibit the dumping of any wastes or other matter including the export of waste to countries for dumping and incineration at sea except for the materials listed in Annex I. (Article 4.1.1, 5 and 6)
Basel Convention	1989	Include plastic waste and microplastics issues into the Basel Convention workstream at COP 13 (Plastic waste in Annex II Y 46 (Household wastes) and Annex VIII (Non-hazardous wastes))
United Nations Environment Programme – Regional Seas Programme and Global Programme of Action Manila Declaration	2003	Regional activities in 12 regional seas
	2012	Prevent marine litter from land-based sources and agree to establish a Global Partnership on Marine Litter (GPML)
G7 Summit	2014	G7 Marine Litter Action Plan
G20 Summit	2017	G20 Marine Litter Action Plan
United Nations Environment Assembly (UNEA) I	2014	Resolutions 1/6: put forward the issue of "Marine plastic debris and microplastics".
UNEA II	2016	Resolutions 2/11: measures to reduce marine plastic litter and microplastics
UNEA III	2017	Resolutions 3/7: combating the spread of marine plastic litter and microplastics.

As the issue of marine pollution including MPs is one of the core agendas of the environmental governance, steps are being initiated not only at global but also at regional and national level.

Different regions of the world are in line to tackle plastic pollution at appropriate regional geographic levels. Regional efforts are those that happen between the national and global approaches [40].

There are multiple regional sea conventions, and actions have been taken in the world, for instance, Regional Action Plan on Marine Litter Management, Northwest Pacific Action Plan, and Marine Strategy Framework Directive adopted by European Union, etc. The EU actively contributes in international efforts to mitigate the plastic pollution including MPs. The EU has set multiple legislative measures and actions plans, such as Directive on Port Reception Facilities for Ship-generated Waste and Cargo Residues (EC2000/59); Directive on Packaging and Packaging Waste (2004/12/EC; EU Bathing Water Directive (76/160/EEC and 2006/7/EC); EC Urban Waste Water Treatment Directive (91/271/EEC and 98/15/EC); EU Environmental Liability Directive (2004/35/EC); and European Strategy on Plastic Waste in the Environment (Green Paper, EC 2013).

At national level, various countries, especially developed ones, have promulgated legislation actions and policy measures to handle MPs issues [16]. Global decision making may direct national incentives, while national-level policy actions are the actual mechanisms for taking concrete actions. Some countries have established laws in order to ban plastic microbeads in both personal care and cosmetic products in order to ensure the stoppage of plastic microbeads into lakes, coastal areas, and the ocean [30].

Table 2: Current regional and national instruments related to MPs [16]

Regional instruments	Time	Specific contents
European Union (EU) Marine Strategy Framework Directive (MSFD) *National instruments (mainly on microbeads in personal care and cosmetic products)*	2008	The first EU legislative instrument related to the protection of marine biodiversity
United States		Honolulu Strategy 2011 Microbead-Free Waters Act 2015 *Cosmetics:* Ban on manufacturing rinse-off products containing plastic microbeads – July 1, 2017 Ban on introduction or delivery of rinse-off products containing plastic microbeads – July 1, 2018 Non-prescription drugs: Ban on manufacturing rinse-off products containing plastic microbeads – July 1, 2018 Ban on introduction or delivery of rinse-off products containing plastic microbeads – July 1, 2019

UK		Plan to ban on rinse-off cosmetics products containing microbeads – 2018; Ban on the manufacture of these products – end of 2017
Sweden		Ban on microbeads in rinse-off products – January 1, 2018
Italy		Ban on microbeads scrub particles in cosmetics – 2020. Ban on plastic cotton buds – 2019.
Canada		SOR/2017-111 Microbeads in Toiletries Regulations *Cosmetics:* Ban on manufacturing and importation of toiletries containing microbeads – January 1, 2018 Ban on the sale of toiletries containing microbeads – July 1, 2018 *Natural health products and/or non-prescription drugs:* Ban on manufacturing and importation of toiletries containing microbeads – July 1, 2018 Ban on the sale of toiletries containing microbeads – July 1, 2019
New Zealand		World Trade Organization notified of draft regulation: ban on microbeads in personal care products – 2018
Australia		Voluntary removal of microbeads – 1 July 2017
South Korea		Ban on the use and sale of plastic microbeads in all cosmetics – July, 2018
Taiwan Province, China		Ban on the manufacture and import of products containing microbeads –January 1, 2018 Ban on sale of products containing microbeads – July 1, 2018

Local communities can also play an important role to address the challenge of MPs. Effective governance can be achieved with effective communication of local communities [41]. It is noted that the tragedy of commons can be governed not only by proper legislation and regulation, but adaptive governance also plays a prominent role. At various points, communities proved to have the capacity for self-management so that they were involved in decision making of resource management [42]. For instance, Bye Bye Plastic Bags (see: *http://www.byebyeplasticbags.org/*) is a successful community-based campaign, which aims to reduce the single use plastics.

The important role of local communities in effective governance is also recognized around the world. For example, proper campaigns are being launched by local administration and Ministry of Climate Change to engage local communities in Islamabad, the capital city of Pakistan, where the use of plastic bags is banned. There are hundreds of groups and local communities engaged in decision making and governance. Local communities in South Korea have used social license to operate reducing the mismanagement of polystyrene buoys [43].

3 Research Methodology

The research methodology depends on the central research question and research objectives of the study. The main objective of this study is to review the evolution of governance steps and policy initiatives to address the issue of plastics pollution with special focus on MPs challenges. The study is conducted by employing literature review as research methodology. Considering the recognition of MPs issues in recent years, focus of this study is set on analyzing plastics governance. The recent research studies and related research papers on plastic governance and plastic pollution are identified through Scopus and Google Scholar for the analysis. The identified studies and related research papers are 78 in number.

4 Discussion and Analysis

Plastic governance is a complex phenomenon due to its transboundary nature and is becoming a serious threat for the world. Due to its complexity, various global, regional, national, and local level countermeasures are being experimented.

At the global scale, there are numerous efforts in the form of conventions and laws to manage the issue of plastic pollution. The legal frameworks are set and demand to carry out actions in order to protect and to preserve the marine environment. These frameworks also emphasize to take concreate steps to control, to reduce, and to prevent pollution of the marine environment from any other source.

Several other activities are also undertaken at international level. For instance, Stockholm and Basel Conventions are important international binding instruments that may provide better opportunities in order to reduce the negative implications of plastics and plastic waste globally. These conventions encouraged to establish public-private cooperation, sharing of best practices, and technical assistance to curb the issue of plastic pollution.

The awareness and scientific knowledge on marine litter and MPs are strongly encouraged at international level. For example, the Joint Group of Experts on the Scientific Aspects of Marine Environmental Protection is responsible to conduct and to support marine environmental assessments, to undertake in-depth studies, analyses, and reviews of specific topics, and to identify emerging issues regarding the state of the marine environment.

The international community is also concerned about the effective implementation of established frameworks and legal aspects. Although many initiatives have been taken at global scale, there is no legal binding as such. A comprehensive report of UNEP says that states are encouraged to *"develop and to implement laws to ban or to diminish the production of single-use trash items and other waste".*

At regional level, seas conventions and action plans on plastic and MPs pollution are there. Regional Seas Program covers 18 regions consisting of over 146 countries that participate in seas conventions and action plans. These seas programs evolved with the passage of time during the previous decades. They have started their work from pollution abatement, but now they have proper systems of monitoring and assessment, land-based and sea-based sources of pollution, oil spill contingency/recovery plans, coastal habitat management, marine litter, and legal and institutional frameworks. Most of these programs act through action plans. The action plans by member governments are adopted to frame a comprehensive strategy and framework in order to protect the environment and to promote sustainable development.

Europe has established four cooperation structures in the form of OSPAR Convention, Helsinki Convention, Barcelona Convention, and Bucharest Convention. The primary objective of these conventions is to protect the marine environment in the member states and in neighboring countries. Their focus is to prevent and to reduce marine litter from both land and sea-based sources by taking a range of actions at national or regional level, such as improved waste and wastewater management, port reception facilities, targeted fishing for litter, education, awareness raising, and outreach activities.

The G20 and G7 also recognize to take actions for prevention and reduction of marine litter in order to preserve human health and marine and coastal ecosystems. G20 encourages actions at regional, national, and local levels to address plastic pollution and marine litter. It also recognizes the role of non-state actors and private sector involvement for resolving plastic pollution. G7 also encourages taking steps to combating marine litter and it stresses the need to manage land- based and sea-based sources, promotion of education, research, and outreach.

At national scale, many developed countries have promulgated legislation and policy actions to address the challenge of MPs. The industries related to plastic in the USA and the UK have implemented "Operation Clean Sweep" to reduce plastic pellet loss to the environment especially during transportation and shipment.

Some European countries like Austria, Belgium, Luxembourg, the Netherlands, and Sweden have accelerated their efforts to ban microbeads from personal-care products in 2015. These countries have shown their concerns that they are harming human health by ending up in seafood. In 2018, Sweden and UK have banned on microbeads in rinse-off products. UK has also launched Beach Cleanup and awareness campaigns like the Green Blue initiative to address the issue of plastic. The Italian parliament asked for ban microbeads particles in cosmetics as of 2020. In addition, Italy has banned plastic cotton buds as of 2019.

A number of non-EU countries such as Canada, New-Zealand, and the USA have banned microbeads or have drawn up voluntary agreements with plastic industry. The developing countries are also in line to mitigate the challenge of plastic pollution and marine litter. In 2002, Bangladesh was the first country to ban plastic bags. Many other countries around the world have taken the same actions. To date, many countries have introduced taxes, bans or restrictions on the use of plastic bags. In some countries such as the USA and Australia the issue is being taken and implemented at a state-based level rather than at national level.

Recently, August 14, 2019, Pakistan banned plastic bags in the capital city of Islamabad, which is also a local level action. Similar measures are observed at many other places in different countries. The local actions are recognized and getting more attentions after the Paris Agreement (PA) that strongly recognizes the role of local governments and non-state actors to deal with climate change.

Mismanaged plastic waste is considered an important source of pollution in the form of waste itself and through degradation into MPs. Recycling is a sustainable and environmental-friendly mechanism, which can transform waste into financial, environmental, and social resources. It does not require advanced disposal technology, which may not be available with developing countries. Therefore, recycling can be an effective approach to reducing the accumulation of plastic waste. National and local governments can opt for adaptation strategies and recycling to solve the problems of waste recycling.

Despite attentions are given to marine plastic pollution, but policy integration and coherence remain a large governance gap at the moment. Multiple positive governance initiatives have been taken but certain challenges for the existing governance mechanisms are identified. These identified challenges are lack of cooperation among transboundary communities, different political interests, weak institutional capacity, and scarcity of financial resources especially in the developing world. The main drivers behind these challenges are also dig out. These identified drivers are regional conflicts, absence of regulatory measures, weak institutional capabilities, lack of intergovernmental cooperation, and absence of engaging local actors in decision making.

The international policy agreements and efforts towards plastics are almost at the same stage to that in which climate change agreements were in 1992. If international efforts come with the same pace as those for climate change, an effective international agreement on plastic pollution can be established in the upcoming decade. It is imperative to take variety of actions, legislative measures, and cooperative approaches will ultimately help to resolve this tragedy of plastic pollution. Science-Policy interface is key for establishing sound and implementable governance initiatives. Education is also important to strengthen community support and understandability of the impact of plastic on the marine environment. Such education and awareness campaigns increase educational and public awareness about plastic pollution, instrumental for adaptive strategies and helpful for the community engagement. Community based actions are needed to incorporate in local and national level action plans. These coordinated and practical actions are helpful to resolve the transboundary problem of plastic pollution and MPs.

5 Conclusions

Microplastics are great environmental transboundary and complex issue. However, the tragedy of the plastic commons is tractable and solvable by incorporating effective policies and coordination to work effectively at global, regional, national, local and individual levels. There are multiple challenges and hindrances for effective plastic governance. Recently, it is recognized one of important areas of research and policy debate but the research and governance mechanisms are still in infancy stage. Most of the countries lack to develop a strategic approach to investigate the primary sources of MPs. Some non-profit organizations and voluntary initiatives are taken place to deal the issue of MPs, but these actions are lacking teeth and commitment to address the major land-based sources of plastic pollution. Another challenge for taking concrete actions is identified as lack of cooperation among transboundary communities due to different political interests. Weak institutional capacity and scarcity of financial resources especially in the developing world are also a contributing factor for weak plastic governance. The main drivers behind these challenges are identified as: regional conflicts, absence of regulatory measures, weak institutional capabilities, lack of intergovernmental cooperation, and absence of engaging local actors in decision making. A societal shift has emerged in the form of a new global social movement advocating awareness of plastic pollution. The PA encourages adaptive form of governance to address the tragedy of commons by increased community engagement and science to guide and to inform actions. Sound regulations and legislation along massive level awareness campaigns related to MPs can play a key role for an effective adaptation governance towards MPs. It is needed to establish binding framework on the same line of climate change to make a sustainable plastic economy possible. Monitoring programs are important for policy interventions, and in prevention and management of MPs pollution.

6 Acknowledgements

The author would like to thank EXCEED Swindon project and DAAD (German Academic Exchange Service) for support to participate at the Regional Workshop on "Microplastics in the Water Environment", 19-21 August 2019, Koh Samui, Thailand.

7 References

[1] R. Geyer, J.R. Jambeck, K.L. Law: Production, use, and fate of all plastics ever made. Science Advances, 2017, 3(7), 1-6.

[2] D.P.J. Kershaw: Marine plastic debris and microplastics–Global lessons and research to inspire action and guide policy change. United Nations Environment Program, Nairobi. (2016) Available at: http://ec.europa.eu/environment/marine/good-environmental-status/descriptor-10/pdf/Marine_plastic_debris_and_microplastic_technical_report_advance_copy.pdf (accessed 15 July 2019).

[3] A.L. Andrady, M.A. Neal: Applications and societal benefits of plastics. Philosophical Transactions of the Royal Society B, Biological Sciences, 2009, 364(1526), 1977–1984.

[4] F. Amélineau, D. Bonnet, O. Heitz, V. Mortreux, A.M.A. Harding, N. Karnovsky, W. Walkusz, J. Fort, D. Grémillet: Microplastic pollution in the Greenland Sea: Background levels and selective contamination of planktivorous diving seabirds. Enviromental Pollution, 2016, 219, 1131–1139.

[5] R.C. Thompson, Y. Olsen, R.P. Mitchell, A. Davis, S.J. Rowland, A.W.G. John, D. McGonigle, A.E. Russell: Lost at sea: where is all the plastic? Science, 2004, 304(5672), 838.

[6] J.A. Van Franeker, C. Blaize, J. Danielsen, K. Fairclough, J. Gollan, N. Guse, P.L. Hanse, M. Heubeck' J.K. Jensen, G.L. Guillou, B. Olsen, K.O. Olsen, J. Pedersen, E.W.M. Stienen, D.M. Turner: Monitoring plastic ingestion by the northern fulmar Fulmarus glacialis in the North Sea. Enviromental Pollution, 2011, 159(10), 2609–2615.

[7] A.K. Baldwin, S.R. Corsi, S.A. Mason. Plastic debris in 29 Great Lakes tributaries: relations to watershed attributes and hydrology. Environmental Science and Technology, 2016, 50(19), 10377–10385.

[8] G. Hetsroni: Particles-turbulence interaction. International Journal of Multiphase Flow, 1989, 15(5), 735–746.

[9] H. Westphalen, A. Abdelrasoul: Challenges and Treatment of Microplastics in Water. Water Challenges of an Urbanizing World. IntechOpen, 2017, 71-83.

[10] J.P.W. Desforges, M. Galbraith, N. Dangerfield, P.S. Ross: Widespread distribution of microplastics in subsurface seawater in the NE Pacific Ocean. Marine Pollution Bulletin, 2014, 79(1–2), 94–99.

[11] J.C. Anderson, B.J. Park, V.P. Palace: Microplastics in aquatic environments: implications for Canadian ecosystems. Enviromental Pollution, 2016, 218, 269–80.

[12] M.A. Browne, S.J. Niven, T.S. Galloway, S.J. Rowland, R.C. Thompson: Microplastic moves pollutants and additives to worms, reducing functions linked to health and biodiversity. Current Biology, 2013, 23(23), 2388–2392.

[13] M.C. Goldstein, M. Rosenberg, L. Cheng: Increased oceanic microplastic debris enhances oviposition in an endemic pelagic insect. Biology Letter, 2012, 8(5), 817–820.

[14] K. Duis, A. Coors: Microplastics in the aquatic and terrestrial environment: sources (with a specific focus on personal care products), fate and effects. Environmental Sciences Europe, 2016, 28(1), 1-25.

[15] C. Zarfl, D. Fleet, E. Fries, F. Galgani, G. Gerdts, G. Hanke. M. Matthies: Microplastics in oceans. Marine Pollution Bulletin, 2011, 62, 1589-1591.

[16] J. Wang, L. Zheng, J. Li: A critical review on the sources and instruments of marine microplastics and prospects on the relevant management in China. Waste Management & Research, 2018, 36(10), 898–911.

[17] I.E. Napper, R.C. Thompson: Release of synthetic microplastic plastic fibres from domestic washing machines: effects of fabric type and washing conditions. Marine Pollution Bulletin, 2016, 112(1–2), 39–45.

[18] M. Filella: Questions of size and numbers in environmental research on microplastics: methodological and conceptual aspects. Enviromental Chemistry, 2015, 12(5), 527–38.

[19] S. Zhao, L. Zhu, D. Li: Microplastic in three urban estuaries, China. Environmental Pollution, 2015, 206, 597–604.

[20] K. Lei, F. Qiao, Q. Liu, Z. Wei, H. Qi, S. Cui, X. Yui, Y. Deng, L. An: Microplastics releasing from personal care and cosmetic products in China. Marine Pollution Bulletin, 2017, 123(1–2),122–126.

[21] S.N. Khan, M. Mohsin: The power of emotional value: Exploring the effects of values on green product consumer choice behavior. Journal of Cleaner Production, 2017, 150, 65-74

[22] P. Bhattacharya: A review on the impacts of microplastic beads used in cosmetics. Acta Biomedica Scientia, 2016, 3(1), 47-52

[23] C. Guerranti, T. Martellini, G. Perra, C. Scopetani, A. Cincinelli: Microplastics in cosmetics: Environmental issues and needs for global bans. Environmental Toxicology and Pharmacology, 2019, 68, 75-79.

[24] S. Klein, I.K. Dimzon, J. Eubeler, T.P. Knepper: Analysis, occurrence, and degradation of microplastics in the aqueous environment. Freshwater Microplastics, 2018, 58, 51–67.

[25] J.A.I. do Sul, M.F. Costa: The present and future of microplastic pollution in the marine environment. Enviromental Pollutution, 2014, 185, 352–364.

[26] M. Cole, P. Lindeque, C. Halsband, T.S. Galloway: Microplastics as contaminants in the marine environment: a review. Marine Pollution Bulletin, 2011, 62(12), 2588–2597.

[27] A.L. Andrady: Microplastics in the marine environment. Marine Pollution Bulletin, 2011, 62(8), 1596–1605.

[28] K. Davidson, S.E. Dudas: Microplastic ingestion by wild and cultured Manila clams (*Venerupis philippinarum*) from Baynes Sound, British Columbia. Archives of Environmental Contamination and Toxicology, 2016, 71(2), 147–156.

[29] C.L. Waller, H.J. Griffiths, C.M. Waluda, S.E. Thorpe, I. Loaiza, B. Moreno, C.O. Pacherres, K.A. Hughes: Microplastics in the Antarctic marine system: an emerging area of research. Science of The Total Environment, 2017, 598, 220–7.

[30] J. Vince, B.D. Hardesty: Plastic pollution challenges in marine and coastal environments: From local to global governance. Restoration Ecology, 2017, 25(1), 123–128.

[31] J. Lee, S. Hong, Y.K. Song, S.H. Hong, Y.C. Jang, M. Jang, N.W. Heo, G.M. Han, M.J. Lee, D. Kang, W.J. Shim: Relationships among the abundances of plastic debris in different size classes on beaches in South Korea. Marine Pollution Bulletin, 2013, 77(1–2), 349–354.

[32] L. Silva-Iñiguez, D.W. Fischer: Quantification and classification of marine litter on the municipal beach of Ensenada, Baja California, Mexico. Marine Pollution Bulletin, 2003, 46(1), 132–138.

[33] M.C. Araújo, M. Costa: An analysis of the riverine contribution to the solid wastes contamination of an isolated beach at the Brazilian Northeast. Management of Environmental Quality, 2007, 18(1), 6–12.

[34] T. Dietz, E. Ostrom, P.C. Stern: The struggle to govern the commons. Science, 2003, 302(5652), 1907–1912.

[35] E.E. Ostrom, T.E. Dietz, N.E. Dolšak, P.C. Stern, S.E. Stonich, E.U. Weber: The drama of the commons. National Academy Press, Atlanta, GA, USA, 2002.

[36] S. Pettipas, M. Bernier, T.R. Walker: A Canadian policy framework to mitigate plastic marine pollution. Marine Policy, 2016, 68, 117–122.

[37] S.B. Borrelle, C.M. Rochman, M. Liboiron, A.L. Bond, A. Lusher, H. Bradshaw, J.F. Provencher: Why we need an international agreement on marine plastic pollution. Proceedings of the National Academy of Sciences, 2017, 114(38), 9994–9997.

[38] M.G. Manea: Human rights and the interregional dialogue between Asia and Europe: ASEAN–EU relations and ASEM. Pacific Review, 2008, 21(3), 369–396.

[39] A. Jetschke: Institutionalizing ASEAN: celebrating Europe through network governance. Cambridge Review of International Affairs, 2009, 22(3), 407–426.

[40] M.A. Browne: Sources and pathways of microplastics to habitats. Marine Anthropogenic Litter, 2015, 229–244.

[41] C.M. Rochman, F. Regan, R.C. Thompson: On the harmonization of methods for measuring the occurrence, fate and effects of microplastics. Analytical Methods, 2017, 9(9), 1324–1325.

[42] B.D. Hardesty, J. Harari, A. Isobe, L. Lebreton, N. Maximenko, J. Potemra, E.V. Sebille, A.D. Vethaak, C. Wilcox :Using numerical model simulations to improve the understanding of micro-plastic distribution and pathways in the marine environment. Frontiers in Marine Science, 2017, 4(30), 1–9.

[43] M. Newson, C. Deegan: Global expectations and their association with corporate social disclosure practices in Australia, Singapore, and South Korea. The International Journal of Accounting, 2002, 37(2), 183–213.

AN EVALUATION OF DOWNSTREAM POLICIES TO REDUCE MARINE PLASTIC LITTER IN THAILAND

Solomon K.M. Huno, Guilberto Borongan, Naoya Tsukamoto

Regional Resource Centre for Asia and the Pacific, Asian Institute of Technology P.O. Box 4, 58 Moo 9, Km. 42, Paholyothin Highway, Klong Luang, Pathum Thani 12120, Thailand;
guilberto@rrcap.ait.ac.th

Keywords: Evaluation, marine plastic litter, plastic waste management policy

Abstract

Marine plastic pollution damages ecosystems and threatens the existence of biodiversity. Thailand ranks among the ten most marine plastic polluted regions in the world. Existing waste management strategies focus on implementation of 3R strategies to manage land-based plastic pollutions that leaks into marine environments. To formulate effective policies to tackle this problem, policies on marine litter must be beneficial to environment and all stakeholders, socially responsible, politically and economically viable, and adaptive enough to remain relevant to the changing waste and technological regimes. Until 2017, Thailand has promoted Municipal Solid Waste Management policies mainly through voluntary policy instruments. However, the launch of the "Plastic Debris Management Plan 2017–2021" introduces a mix of policy measures including direct regulatory and market instruments to drive policy implementation. An evaluation framework was applied to analyze and to understand the political, economic, social, legal and technological dimensions of the existing downstream marine plastic management policies.

1 Introduction

Marine plastic pollution has garnered significant global attention due to the visible impacts on aquatic life and their persistence in the environment. Plastics and microplastics (MPs) in the aquatic environment are considered a man-made hazard to various ecosystems with high-risk impacts to fauna therein. The impacts of marine plastic pollution on aquatic life are evident from the deaths to animals from entanglement with macroplastics plastic, which impairs the normal biological functioning of the animal body [1]. Further, plastic litter ingested may also be a source of potentially adverse toxic effects on marine organisms. Southeast Asian development process has been environment-intensive and as such raised concerns for increased attention to emerged environmental problems that constrain sustainable growth [2]. Previous studies noted the complex relationships between environmental impacts and or environmental degradation rate and economic growth. These relationships albeit a complex one, have been shown to have close associations with the phase of economic development of the country as well as other contextual macro-economic factors [3].

Thailand, like other ASEAN Member States has experienced rapid infrastructural development and economic growth with the industrial sector and domestic consumption as the main stay of the country, boosting the economy [4, 5]. The linkages between plastic waste generation and economic growth can be drawn from the understanding of waste as a byproduct of economic activities of business enterprises, households and government, having implications for productivity, expenditure and the environment. Consumer decisions and preferences in demanding goods and services also has implications on plastic waste compositions and volume of total waste generated, which has an implied effect on government spending on waste management. Rapid economic growth in Thailand over the past decade has come at a cost to the environment, mainly regarding increasing municipal solid waste generation from urbanization, industrial waste and also from domestic consumption [6-9].

South East Asia accounts for almost 20% of global plastics consumption, which is expected to continuously grow. The dominant source of marine plastic litter comes from leakages from mismanaged land-based plastic waste [10]. With an estimated total coastal area of 350,682 km^2, comprising 35,834 km^2 land area and 314,845 km^2 of marine area, Thailand has a total coastline of 3,148 km that stretches with 2,055 km along the Gulf of Thailand covering 17 provinces, and 1,093 km along the Andaman Sea covering 6 provinces [11]. Thus, the management of plastic waste in Thailand and other countries within the Lower Mekong Region is significant in stemming the flow of mismanaged land-based plastic waste transport into the ocean [10]. Plastic litter constitutes the highest fraction of coastal debris in many beaches of the country [12, 13]. Impacts of macroplastic and MPs on marine organisms in the Gulf of Thailand have been identified [14]. Plastic consumption is an integral part of Thai consumption culture, where plastic packaging is the most dominant packaging for fast foods. Thus, the proper end-of-life treatment or management of plastic waste in MSW requires the appropriate interventions to limit leakage into water environment. Plastic is mostly utilized as packaging and electronics, consumer and institutional products (e.g., dinner and kitchenware, toys and sporting goods), and textiles, among others.

Notwithstanding, the economic benefits of the plastic industry at national, regional and global level, plastic leaked into the environment at various stages along the value chain (i.e., manufacturing, processing, transport and product use, disposal and recycling). While plastics leakage into marine environment may occur as MPs or macroplastic losses, the latter has often been attributed to mismanaged municipal solid waste [15]. The effects of marine plastic and poor postconsumer management of plastic products have led to challenges in the increase in plastic litter into the oceans. Increasing demand for plastic has been attributed to the versatility of plastic application in a wide range of consumer products manufacture with packaging (30%), building and construction (17%), and transportation (14%) having the largest product share. The large market share of plastic packaging has remained high in most developing countries largely due to the relative convenience of access, ease of use and affordability, particularly the single use packaging. Polyvinylchloride (PVC), polypropylene (PP), low density polyethylene (LDPE) and linear low-density polyethylene LLDPE), high density polyethylene (HDPE) and polyethylene terephthalate (PET) are among the most widely used, about 50% of total plastics usage [16]. Plastic waste is collected as part of municipal solid waste (MSW) together with other streams of waste. Notably, MSW generation has increased from an estimated 23.93 million to/yr in 2008 to an estimated 27.82 million t/yr (= 74,998 t/d) in 2017 (equivalent to 1.13 kg/d per person) [17].

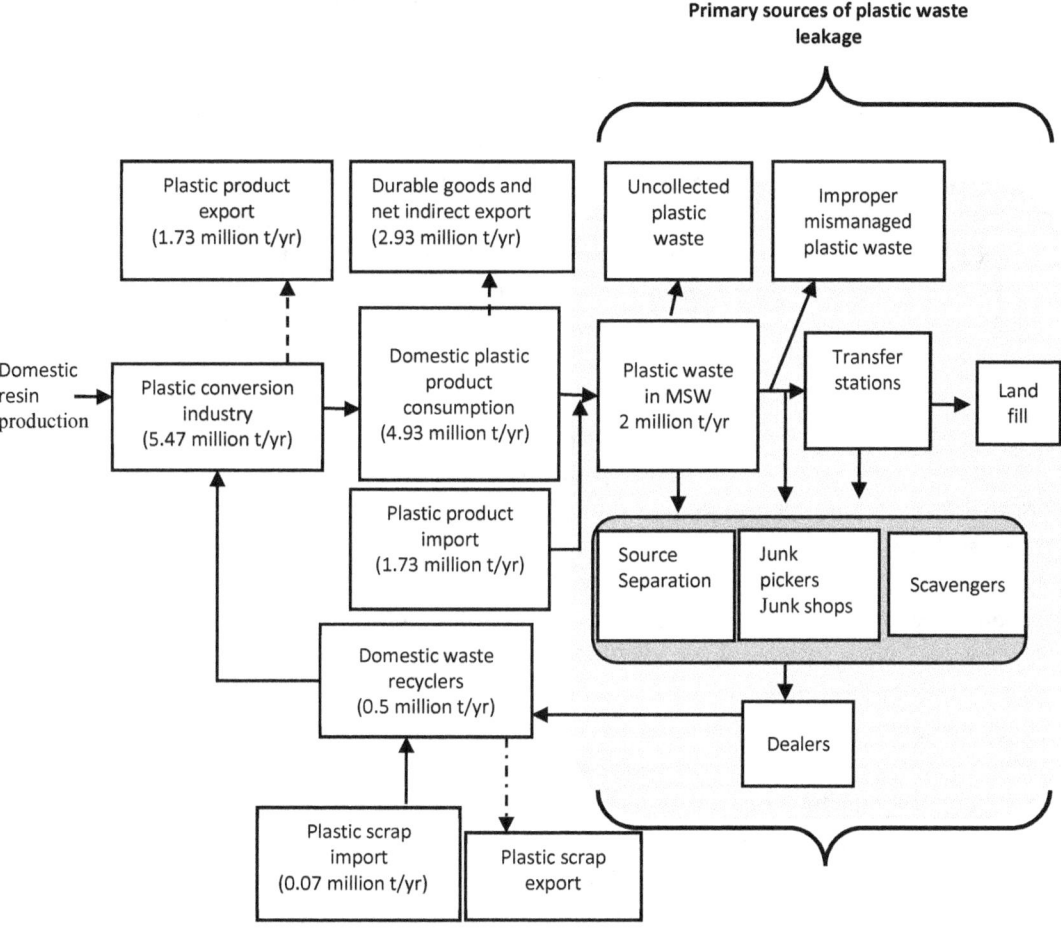

Figure 1: Plastic flow in Thailand showing the major sources of mismanaged plastic leakage at the downstream of the flow chain *(developed by authors, data from Thailand Plastic Institute)*

In 2018, about 2 million tons of plastic waste was produced as part of MSW (Figure 1). Recyclable waste comprising mainly of PET bottles accounts for about one-third (500,000 tons), while the rest of 1.5 million tons is plastic waste, plastic bags, other plastic garbage, including glass, boxes, trays, bottles, and bottles with lids [18]. Notably, about 0.03 million tons out of the improperly disposed solid waste (7.36 million tons) was unmanaged across the country [19, 20]. Plastic waste constituted about 20.9% of total waste generated in Thailand in 2017 and has shown an annual increase of 2 million tons since the past decade [21].

The governance of waste management in Thailand is carried out within the decentralized governance framework, where environmental management has become a main function of local authorities. The central government through the Ministry of Natural Resources and the Environment (MONRE, ONEP, DEQP and the PCD) through inter-ministerial collaborations with other line ministries, such as the Ministry of Interior (Local Authority), the Ministry of Public, Health, the Ministry of Energy and the Ministry of Industry (Department of Industrial Works) performs the policy development, regulatory and supervisory functions. Thailand's political and administrative structure has the provincial administration organizations (PAO) except Bangkok and Pattaya as the largest of the local government organizations. The PAO's manage and provide public services within their command area, facilitating the works of municipalities and the sub-district administrations effectively through collaborations with other administrations within the same province. The Department of Local Administration under the Ministry of Interior consults, supports and facilitates the Local Government Organizations (LAOs) to efficiently manage and to provide public services according to their roles and functions under the principle of good governance.

Though, administrative power has been devolved to the local level, having districts and sub-district administration organizations as the lowest administrative units. The local powers are under the state powers; the local administrations are not independent bodies, but they are under the national laws, set up for the benefit and well-being of the members of the community. The responsibilities of the development of policy recommendations on national policy on waste management and minimization, development of standards and guidelines, including provision of technical support, monitoring and reporting, compliance investigation falls within the purview of the Pollution Control Department (PCD). The LAOs are main implementers of national waste policies and are directly involved in ensuring efficient and effective collection and disposal by exhibiting good governance through effective policy implementations and high degree of independence in financing and managing public services. Thailand has shown great commitment to improving environmental performances by the enacting various waste management policies and strategies to ensure. The development of regulations, policy, action plan and information products regarding waste management prescribed under the Act B.E. 2545 (2002) are under the purview of the PCD. Table 1 shows a list of related MSW management polices related to plastic waste management.

Table 1: Policies related to plastic waste management

General municipal solid waste policies	Voluntary Instrument	Direct regulatory instrument
The National Economic and Social Development Plan No. 12	✓	
The National Environmental Basic Plan	✓	
National Solid Waste Management Master Plan (2016 - 2021)	✓	
The National 3R Strategy	✓	
20-Year Pollution Management Strategy and Pollution Management Plan 2017-2021.		✓

Action Plan "Thailand Zero Waste" (2016 – 2017)	✓	
Strategic Plan on Packaging and Packaging Waste Management (Draft)		
Action Plan "Clean Province" (2018)	✓	
Plastic waste specific policies		
Plastic Debris Management Plan 2017-2021	✓	✓
National Roadmap for the Development of Bioplastics Industry	✓	✓
Roadmap on plastic waste management 2018-2030	✓	✓

However, most measures are taken in the form of voluntary campaigns, which are not enforced in terms of legal measures. The Ministry of Interior and Local Authorities are operators implementing these policies. Waste management efforts by the country has been focused achieving maximum impact by leveraging synergies from simultaneous promotion of environmental policies focusing on environmental research on plastic waste, intensifying public education and awareness, enactment of sound policies, and conducting best practices and 3R strategies. Unfortunately, policy interventions employed by the Thai governments to reduce generation of plastic waste largely remains voluntary measures [22]. The country is faced with challenges of providing adequate waste collection bins, poor source separation, low coverage of waste collection and waste transport services, inadequate infrastructure for waste treatment and disposal, illegal waste dumping, open burning, among others. To address these challenges, efforts have concerted by government and international bodies towards reducing the leakage of land-based plastic waste into marine environment. Notwithstanding, the challenges of land-based plastic waste leakage into marine environment persists. Policy interventions implemented over the years have appeared to be ineffective in reducing plastic waste and stem land-based plastic waste leakage into marine environment. A range of governance gaps or barriers have been suggested as reasons. Significantly, the implementation of these policies has produced marginal results, with most policy targets not achieved.

Hitherto, government policy actions have not been specific towards plastic waste but as an integrated approach towards MSW management. This approach agglomerated plastic waste as part of other MSW streams. However, since 2018, specific actions to abate plastic waste generation and subsequent leakage have been initiated by the government. To understand the effectiveness of the menu of policies implemented over the past years, an *ex-post* evaluation of the various policies will provide a good understanding feedback to policy makers. However, that often is not the case; responsible authorities are often constrained by limited economic resources to carry out systematic evaluation of environmental policies. Clearly, environmental policy interventions are implemented in combination with other policy interventions and thus, disentangling their specific contribution becomes a complex, and sometimes impossible to evaluate. Other significant barriers also involve the lack of adequate data necessary to perform evaluations and institutional obstacles arising from responsibility overlaps by government departments. Notwithstanding, a systematic evaluation provides feedback, which can help to improve the design and performance of environmental policies.

This study, therefore, aims to apply comprehensive *ex-ante* evaluation approaches to examine the content of implementation processes of such policies, adopting suitable policy analysis tools to asses and to evaluate the contribution of simultaneously implemented portfolio of waste management strategies/policies' contribution to reducing marine plastic litter in Thailand.

2 Materials and Methods

The study examines the likely effectiveness of a portfolio of downstream environmental policy measures, implemented by the Thai government and their contributions to the reduction of marine plastic liter. The analysis took cognizance of the general inclusion of plastic waste as a fraction MSW (including household and industrial waste), which has not been considered separately. Hence, waste management policy instruments were not very explicit as plastic waste targeted in nature. Also, it was observed that majority of waste management policies are integrated in time and scope, with limited monitoring data. As such, to avoid the jeopardy of indeterminacy and to reduce the intricate complexity of likely empirical relations and the associated vast information requirements, an *ex-ante* policy evaluation approach was used to analyze the merits, sustainability and contextual relevance of the policy interventions to plastic waste generation and leakage reduction to marine environment. The implications of an alternative no action or business as usual scenario by the Thai government was also considered. Thus, a general level policy evaluation is presented with conditions and for their application conditions to draw a picture of the effectiveness of policy instruments as well as the unanticipated effects of the environmental policy instrument(s) on marine plastic litter.

2.1 Data acquisition
Preliminary dataset for this study was acquired during the implementation of a donor-funded project undertaken by the authors. Published reports on plastic waste management from government agencies and organization, international development organizations, development finance institutions, conference and workshop proceeding of marine plastic litters. Further, peer reviewed journal publications were systematically searched in Web of Science and Scopus on using the key word combinations; *"Thailand and waste management policies"* and *"Marine plastic litter and Thailand and waste management policies"*. Recent published reports and conference proceeding reports on plastic waste management policies were reviewed.

Quantitative and qualitative data on land-based plastic and marine plastic litter from secondary sources were curated and subsequently validated to eliminate and to correct internal inconsistencies. Disaggregated waste generation data comprising plastic waste, waste utilization, recycling rate, mismanaged waste, plastic waste leakage, total plastic consumption, government investment in plastic waste management and related policy documents was sourced from published reports from government organizations such as the Pollution Control Department (PCD) of the Ministry of Natural Resources and Environments (MONRE), Import-Export Data (Comtrate), Department of Industrial Work (DIW) and Ministry of Energy (MOE) and Thailand Board of Investment. Macro- and socio-economic indicator data were sourced from various government organization publications including Thailand Bureau of the Budget, Board of Investment, the World Bank group and ASEAN statistical yearbook.

2.2 Data analysis and policy content evaluation

Environmental policies related to plastic waste and marine litter management implemented over the past two decades were reviewed. A systematic content analysis of policy characteristics covering the policy objectives, policy actions, policy instruments and implementations plans were analyzed to capture and to draw insights on the likely factors that influenced the achievement of the various policy objectives. Expert judgements about the enterprise environmental conditions of the policy legitimacy, implementation and uptake within the existing governance framework of waste management in the country were also evaluated. Policy implementation strategies were also evaluated considering context specific challenges faced by the country using a PESTEL (political, economic, social, technological environmental and legal) multi-dimensional analytical framework. Policy characteristics were evaluated using the evaluation criteria shown in Table 2.

Table 2: Criteria for analyzing of down-stream polices

Category	Criteria
Policy background	- Baseline studies conducted - Policy rationale is evidence informed - Background information accessible
Goals and objectives	- Policy goals and objectives are clearly stated - Clear indicators set with concrete goals - Policy action plan and clear implementation roadmap stipulated
Policy Legitimacy and feasibility	- Legitimacy approach - Social and political acceptability - Public support for policy - Legal framework for implementation
Policy instrument and mechanisms	- Political-administrative instruments - Regulatory instruments - Social instruments
Actor relations	- Who influences whom? - What is the influencing mechanism? - What is being influenced? - What is the operational goal?
Administrative Capabilities	- Within administrative capacities to implement policies

Monitoring and evaluation	• Monitoring and evaluation plans clearly designed at marked stages of implementation • The outcome measures are identified for each objective/goal • The data collected for evaluation collected before, during, and after the introduction of the new • Feedback loop from policy evaluation
Multi-dimensional feasibility of strategy	• Policy fits in the broader institutional framework of society; Political Economic Social, Technological Environmental and Legal

3 Results and Discussion

Thailand's commitment to improving the country's environmental performance has been demonstrated through the implementation of various waste management policies. As a result, Thailand's Environmental Performance Index (EPI) ranks 121 out of 180 countries surveyed. Within the Lower Mekong Region, Thailand ranks ahead of Vietnam, Laos PDR and Myanmar in the year 2018 [23]. Regardless of the country's has missed out on major policy targets aimed at improving solid waste management and essentially to reduce plastic pollution and land-based leakage to water environment. Investment in waste management collection and disposal is needed to avoid illegal disposal of plastic waste, which is prevalent across the country (Figure 2). Plastic waste collection and disposal remain the primary source of plastic waste leakage into water environment.

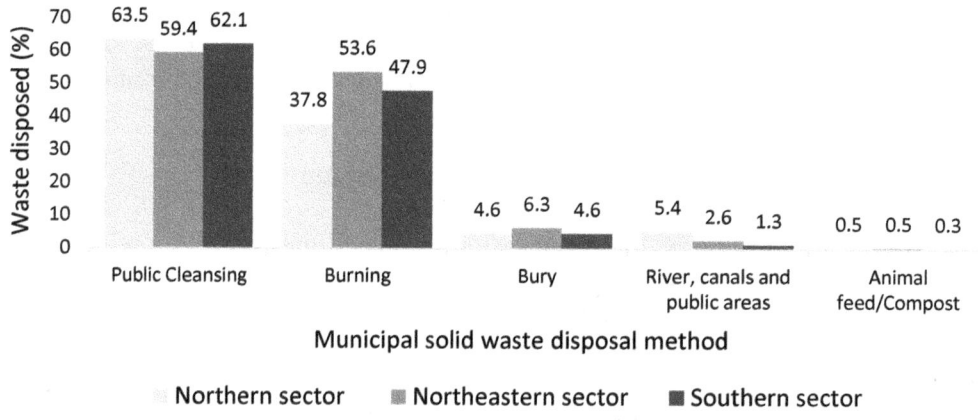

Figure 2: Household waste disposal survey from 100 household samples across Thailand
(Adapted from Thailand Statistical Yearbook (2018); Data for 2017)

3.1 Policy actions to prevent marine plastic litter
3.1.1 Policy background, goals and objectives
Majority of related policies developed by the ONEP/PCD, MONRE is carried through research and consultation where necessary, with environmental networks comprising universities, relevant

private sectors, Department of Environmental Quality and Promotion (DEQP) and stakeholders from communities. However, baseline studies prior to the formulation of various policies have not been published. Nevertheless, background information and rationale for policy formulations is openly publicized and accessible. Baseline studies highlighting the existing situation for targeted interventions provide valuable information to analyze, to monitor, to measure and to evaluate policy impacts. Essentially, critical areas of focus by policy interventions in preventing plastic waste leakage into water environment can be understood by analyzing material flow, hotspot analysis, value flow analysis, value chain, and cost models. Policy goals of each policy were clearly stipulated, and indicators stated policy goals and objectives, including indicators and action plan, and clear implementation roadmap stipulated are common to all the formulated policies, albeit some appear ambitious. Development of policy documents through step by step analysis of situation is as critical to the policy development process as the implementation success of the policy. This process maps out the goals, opportunities, obligations and resources to be recognized as well as the policies adherence to relevant principles and appropriate placement within stakeholder and legislative support framework [24].

3.1.2 Policy legitimacy and feasibility
Apart from judging policy success in terms of effectiveness, efficiency, relevance, equity and feasibility, policy legitimacy is crucial for the successful implementation of any policy. A policy that fails to garner enough support from the public and relevant stakeholders may be unsuccessful in achieving its goals [25]. Similarly, legitimacy can offer great leverage to authorities for stakeholder support in policy implementation, costs reductions associated with monitoring and enforcing public compliance [26, 27]. Existing land-based plastic waste management policies adopted output-oriented legitimacy design processes, which offer a rather quick process of policy formulation as opposed to a slower provides input-oriented is citizen-centered and participatory and democratic [28]. Though, the policy formulation process included consultation with relevant institutions and government organizations. The inclusion of the public has the potential to increase the implementation and citizenry compliance. The formulation of the Plastic Waste Management Roadmap had limited consultations with institutional actors such as the Federation of Thai Industries (FTI), plastic waste management sub-committees, international development organizations, private development organizations and public sector over a total of about 11 working group meetings. This essentially downplays the role of the public consumers, who play major part in the compliance and successful achievements of policy outcomes.

3.1.3 Policy instrument and mechanisms
Waste management policies have generally been adopted voluntary (cultural/informational) policy instruments to manage plastic waste. Thailand has mostly tackled plastic waste management by promoting 3R activities nationwide. As a voluntary and/or informational instrument, actors were not obliged to comply. Though this has achieved some level of success by increasing the recycling rate to about 25%, there is still significant loss of plastic waste along the plastic waste value chain. However, in response to the fast-rising concerns to marine plastic litter, the Plastic Waste Management Roadmap stipulated the adoption of a direct regulation and prohibition instrument,

which saw the ban or discontinue the use of seven plastic items. While this appears laudable, the enforcement of policies of regulatory and economic measures such as tax and levies on single use plastic bags, encouraging product stewardship through extended producer responsibility (EPR) models would contribute significantly to reduction of plastic waste leakage to marine environment. Current plastic waste management policy intends to adopt an integrated mechanism to drive the policy implementation. These include the use of behavioral change, enactment of laws, economic instruments and budgeting tools that achieve policy outcomes.

3.1.4 Actor relations

The various actors involved in the implementation of plastic waste management specific policies can be classified as government actors - local government organizations (LAO's) influenced by the central government agencies, who formulate policies for implementation. The top down influence mostly inhibits successful implementation of policies by LAOs, who often are laden with budgetary constraints and capacity to implement effectively. A second form of influence involves government actors including central government and local government organizations on one side and private firms and non-governmental actors on the other. This influence seeks to court the support of private sector in the implementation of the policies. Examples of private firms include the Federation of Thai Industries, development partners, etc. The third form is the private firm to firm influence for market forces and dynamics to remain competitive and socially responsible to consumers.

3.1.5 Administrative capacities at local government level

Local governments that are responsible for implementation of all waste management policies including plastic waste policies within their command areas under the current waste management directives are faced with significant challenges. These challenges include the lack of technical and financial leverage to be able to actualize central government directives. The situation is severe for the local governments in rural areas. Apart from that the inadequacy of subventions from the central government is woefully inadequate, the budgets are heavily constrained, and thus, environmental considerations are placed secondary to priority needs such as promotion of vocational employment, agriculture and Infrastructure development, with environment being an average priority area for investment. Also, systemic challenges inherent with the local government governance structure may be blighted by certain level of administrative sectionalism, which impacts policy uptake and implementation.

3.1.6 Monitoring and evaluation

Monitoring and evaluation of enacted policies objects are very crucial for realizing the policy goals. However, there appears inadequate monitoring and evaluation support to policy implementation. These can be partly attributed to the huge funding needs for the government to undertake such periodic exercises.

3.2 Multi-dimensional policy space for effective plastic waste management policy implementation

The effective implementation of plastic waste management policies in Thailand requires consideration for multifactor feasibility of this range of policies. The multi-dimensional factors inform the adaptive capacity of policies with the plastic waste leakage into water and marine environment.

3.2.1 Political factors

Effective implementation of waste management policy is largely affected by the political setting at the local government level, local administrative level central and national level. Primarily, political stability has implications for policy continuity as well as the resilience of the sector. The inherent political structure of the local level governance, bureaucratic interferences, corruption and political soft controls on waste management must be overcome to ensure little to no interference in policy implementation. Considering the huge shortfalls in investment in the waste management sector of Thailand, private sector participation in various aspects of the plastic waste management value chain will boost the potential for improving waste collection and recovery of recyclable wastes, which otherwise may end in landfills or marine environments. Private sector participation requires strong political will to pass favorable policies that promote a win-win private public partnership with local government or organizations. Tax rates, tariffs incentives, pricing regulations of recycled products as well as anti-trust laws related to plastic waste management must stimulate private sector participation in the sector.

3.2.3 Socio-economic factors

Implementation of land-based plastic waste management policy is particularly affected. While certain policy tools may be well applicable for certain parts of the country, other parts, particularly the rural areas, may fall short to support the necessary revenue generation models for the local LAOs. Waste generation and waste compositions of the various provinces and LAOs are largely influenced by the socio-economic factors as well as geographic locations. Apart from the disparities in the set minimum wages of the various provinces [26], the southern coastal parts the country are major tourist destinations posting as potential hotspot areas for marine plastic litter. As such, plastic waste policies must be adaptable and context specific in their approach for maximum impact. Blanket type policy approach often proves to be less effective in reducing mismanaged plastic and land-based leakage into the ocean. The viability of waste management services, formal and informal recyclable plastic recovery business hinges on the labor costs, regulation and tax regimes, promote competitiveness, productivity and economic vibrancy of the province or district [27]. Notwithstanding adopting a bottom-up approach for adaptive policies initiated and owned by community leaderships with a community wide support and commitment and government agencies and private sector has great potentials to effect change.

3.2.5 Techno-environmental factors

Technology development and evolution precedes environmental policies evolution mainly due to the slow policy formulation processes. A firm's analysis of available waste management technologies and the rate of development have potentials to impact on cost structure and technological diffusion levels on waste collection, processing, utilization and disposal. Policy options in the downstream could be of the industry but also the speed, at which technology disrupts that industry. Slow speed will give more time, while fast speed of technological disruption may give a firm little time, to cope and to be profitable may be limited by lax patent law in the country. An example is the legislation on regulating the use of plastic in food packaging, which has implications for the recycling of certain plastic waste, thereby limiting the product and value offerings that plastic recycling could provide.

3.2.7 Legal factors that impact waste management
The legal dimensions of plastic waste management policies explore the robustness of the legal framework and institutions to support the implementations of enacted policies. These involve the capacity to enforce as well as to prosecute infractions. However, weak legal systems lack the power to exact appropriate punishments. For example, the violation of transboundary movement and importation of low value plastic waste, which end up in landfills and open waste dumps in contravention of the law, must be prosecuted to serve as deterrent. The implementation of various waste policies such as the current ban on the seven plastic types in the country requires strong legal support to prosecute violators and smugglers from neighboring countries. The gradual introduction of direct regulatory policy instruments to replace the voluntary measures close and clear laws or legal reforms must be done to define civic duties for promoting plastic waste, and recycling waste separation at source will be needed to guide the implementation.

4 Conclusions

Until 2016, Thailand's downstream policies related to plastic waste management has largely been developed and implemented through voluntary policy measures. These measures rely mainly on public awareness programs on 3R activities to reduce plastic waste. Policy formulation processes largely was done through the top down approach albeit via a consultative process. These policies have yielded limited results in reducing mismanaged plastic litter. Adequate waste collection services are lacking and thus result in plastic waste leakage into the aquatic environment. Underinvestment in waste management sector severely affects the implementation of objectives by the local government organizations (LAOs). Some LAO's lack adequate technical and financial supports to implement appropriate plastic waste management systems put in place that can reduce the environmental impact and help creating a sustainable, sound material-cycle society. The implementation of downstream policies initiated at local levels promises to be effective through increased budgetary support from central and provincial governments. Policy frameworks in place to solve the plastic waste problem using the regional waste management framework must be carried out in consonance with the existing political, socio-economic, techno-environmental factors and legal contexts, which are specific to each local area.

5 Acknowledgements

The authors would like to thank DAAD and the project Exceed Swindon to support their participation at this workshop.

6 References

[1]   Carlos de Sá, L., Oliveira, M., Ribeiro, F., Rocha, T.P, Futter, M.N.: Studies of the effects of microplastics on aquatic organisms: What do we know and where should we focus our efforts in the future? Science of The Total Environment, 2018, 645, 10259-1039.

[2]   Hamdan, R., Ab-Rahim, R., Fah, S.S.: Financial Development and Environmental Degradation in ASEAN-5. International Journal of Academic Research in Business and Social Sciences, 2018, 8(12), 14–32.

[3]   Bureecam, C., Chaisomphob, T., Sungsomboon, P.: Material flows analysis of Plastic in Thailand. Thermal Science, 2018, 22(6A), 2379-2388

[4]   ASEAN Economic Integration Brief (AEIB): ASEAN Economic Integration Brief No. 04 / November 2018, Community Relations Division (CRD) of the ASEAN Secretariat, ASEAN. Jakarta. 2018. (*Accessed July 12, 2019*) from https://asean.org/storage/2018/11/AEIB_4th-Issue_r1.pdf

[4]   OECD: Economic Outlook for Southeast Asia, China and India 2018: Fostering Growth through Digitalization. OECD Publishing, Paris (2018),

[5]   Rigg, J.: Counting the Costs: Economic Growth and Environmental Change in Thailand. Institute of Southeast Asian Studies, Thailand, 2000; ISBN: 978-981-3055-08-7

[6]   Khajuria, A., Yamamoto, Y., Morioka, T.: Estimation of municipal solid waste generation and landfill area in Asian developing countries. Journal of Environmental Biology, 2010, 31(5) 649-654.

[7]   Sjöström, M., Östblom, G.: Decoupling waste generation from economic growth - A CGE analysis of the Swedish case. Ecological Economics, 2010, 69 (7), 1545–1552.

[8]   Chen, Y.: Effects of urbanization on municipal solid waste composition. Waste Management, 2018, 79, 828-836

[9]   Lebreton, L.C.M., van der Zwet, J., Damsteeg, J., Slat, B., Andrady, A., Julia Reisser, J.: River plastic emissions to the world's oceans. Nature Communications, 2017, 8, 15611

[10] Prabpriree, M.: The Royal Coast Tourism: Area Potential and Integrated Management for Sustainable Tourism Development. ASEAN Community Knowledge Networks for the Economy, Society, Culture, and Environmental Stability 2015, 3(2), 465-477.

[11] Thushari, G.G.N., Chavanich, S., Yakupitiyage, A.: Coastal debris analysis in beaches of Chonburi Province, eastern Thailand as implications for coastal conservation. Marine Pollution Bulletin, 2017, 116, 121-129.

[12] Ballesteros, L. V., Matthews, J.L., Hoeksema, B.W.: Pollution and coral damage caused by derelict fishing gear on coral reefs around Koh Tao, Gulf of Thailand. Marine Pollution Bulletin, 2018, 135, 1107-1116.

[13] Azad, S.M.O., Towatana, P., Pradit, S., Patricia, B.G., Hue, H.T.T.: Ingestion of microplastics by some commercial fishes in the lower Gulf of Thailand: a preliminary approach to ocean conservation. International Journal of Agricultural Technology, 2018, 14(7), 1017-1032.

[14] Boucher, J., Friot, D.: Primary Microplastics in the Oceans: a Global Evaluation of Sources. International Union for Conservation of Nature and Natural Resources, Gland, Switzerland, 2017

[15] United Nations Environment Programme (UNEP) DTIE (Division of Technology, Industry and Economics) IETC (International Environmental Technology Centre): Converting Waste Plastics into a Resource: Assessment Guidelines. United Nations Environmental Programme, Division of Technology, Industry and Economics, International Environmental Technology Centre, Osaka/Shiga, Japan, (2009).

[16] Pollution Control Department: Ministry of Natural Resources and Environment. Booklet of Thailand State of Pollution, 2017, Bangkok, PCD, 2018

[17] United Nations Center for Regional Development (UNCRD): Ninth Regional 3R Forum in Asia and the Pacific, "3R as a way for moving towards sufficiency economy – Implications for SDGs", 4-6 March 2019, Bangkok, PCD, 2019.

[18] Pollution Control Department: Ministry of Natural Resources and Environment. Booklet of Thailand State of Pollution, 2018, Bangkok, PCD, 2019

[19] Wichai-utcha N., Chavalparit, O.: 3Rs Policy and plastic waste management in Thailand. Journal of Material Cycles and Waste Management, 2019, 21(1), 10-22.

[20] Tangwanichagapong, S., Nitivattananon, V., Mohanty, B., Visvanathan, C.: Greening of a campus through waste management initiatives. International Journal of Sustainable Higher Education, 2017, 18(2), 203–217.

[21] Huang, B., Xu, Y.: Environmental Performance in Asia: Overview, Drivers, and Policy Implications. ADBI Working Paper 990, Asian Development Bank Institute, Tokyo, 2019

[22] Cheung K.K., Mirzaei, M, Leeder, S.: Health policy analysis: a tool to evaluate in policy documents the alignment between policy statements and intended outcomes. Australian Health Reviews, 2010, 34, 405-413

[22] Park, C., Lee, J., Chung, C.: Is "legitimized" policy always successful? Policy legitimacy and cultural policy in Korea. Policy Sciences, 2015, 48, 319-338.

[23] Buchanan, A.: Justice, Legitimacy and Self-Determination: Moral Foundations for International Law. Oxford: Oxford University Press, UK, 2003.

[24] Jagers, S. Simon Matti, S., Nordblom, K.: Nordblom. QoG Working Paper Series 2016, 15, University of Gothenburg, Göteborg; ISSN 1653-8919

[25] Montpetit, E.: Policy design for legitimacy. Public Administration, 2008, 86(1), 259–277.

[26] Ministry of Labor: Thailand announces new minimum wage rates. Ministry of Labor, Bangkok, 2018;
(Accessed October 14, 2019) https://www.boi.go.th/index.php?page=demographic

[27] Pumpinyo, S., Nitivattananon, V.: Investigation of barriers and factors affecting the reverse logistics of waste management practice: a case study in Thailand. Sustainability, 2014, 6, 7048-7062.

www.ingramcontent.com/pod-product-compliance
Lightning Source LLC
Chambersburg PA
CBHW080400030426
42334CB00024B/2940